Carbon Monoxide
A Clear and Present Danger

Third Edition

Dwyer / Leatherman / Manclark / Kimball / Rasmussen

ESCO Press
A Division of
Educational Standards Corporation
www.escoinst.com

Copyright © 2003
ESCO Press

All rights reserved. Except as permitted under
The United States Copyright Act of 1976, no part
of this publication may be reproduced or distributed
In any form or means, or stored in a database
or retrieval system, without the prior written
permission of the publisher, **ESCO Press**.

ISBN 1-930044-20-8

Disclaimer

This book was written as a general guide. The authors and publishers have neither liability nor can they be responsible to any person or entity for any misunderstanding, misuse, or misapplication that would cause loss or damage of any kind, including loss of rights, material, or personal injury alleged to be caused directly or indirectly by the information contained in this book.

Printed in the United States of America
7654321

Table of Contents

Introduction	i
Important Pressure Measurements and Conversions	iv
Section 1: Carbon Monoxide	
Carbon Monoxide	1
Signs and Symptoms of Carbon Monoxide Poisoning	5
Health Effects of Carbon Monoxide Poisoning	6
Study: Carbon Monoxide, Invisible Destroyer of Health and Life	10
Sources of Carbon Monoxide Poisoning	24
Understand How Your Test Instruments Calibrate	26
CO Alarms are Warning Devices	27
What Type of CO Alarm Should be Installed?	28
Changes in the UL Listing for Carbon Monoxide Alarms	30
Why Wait for the Alarm or Injury?	34
Code Compliance	36
Documentation	38
Responding to a Carbon Monoxide Alarm	40
First Response	41
How Much Carbon Monoxide is too Much?	45
CO Air Free Standard	47
Carbon Monoxide and Combustion Testing Procedures	49
Section 2: Combustion	
Combustion	55
Principles of Combustible Gas for Technicians	58
Heating Value	62
Controlled Gas Fuel and Combustion	63
Control Fuel Gas	64
Clocking a Gas Meter	67
Advantages of Measuring O_2 vs CO_2	69
Relationship Between O_2, CO_2 and Excess Air	71
Oil Fired Burners	73
Fuel Delivery, Air, Combustion, By-Product Production	74
Types of Efficiencies	77
Burner Operation	78

Table of Contents

Section 2 continued:

Time, Temperature, Turbulence	80
Time	80
Temperature	83
Turbulence	84
Oxides of Nitrogen (NOx)	85
Draft	88
Generally Acceptable Draft Measurements	91
Combustion Testing Procedures	95
Oxygen, Carbon Monoxide and Stack Temperature	96
Smoke Testing	102
Acceptable Combustion Test Results	103
Accurate Testing	104
Proper Venting	105
Combustion Air	105
Make Up Air	105
Thermal Shock	105
Boilers	107
Stack Temperature	109
Modulating Burner Tune-up	110
Suggested Procedure for Setting up a Modulating Gas Fired Power Burner	113
Savings Potential	116
Combustion Troubleshooting Guide	118

Section 3: Pressure Measurements

Pressure Measurements for Buildings HVAC Installation, Service and Maintenance	123
Reasons Why More Quality Assurance Pressure Testing is Needed	124
Societal Trends that will Increase Quality Assurance Pressure Testing	125
What is Pressure?	126
Warmer Dense Air	127
Air Flow by Building Design	129
Types of Manometers	130
The Manometer Law	131
Duct Pressures	132
Pitot Tube	133
Data Plates	134

Table of Contents

Section 3 continued:

Combination Gas Control	135
External Static Pressure Testing	136
Pressure Drop Across an Air Filter	137
Measuring Airflow by Device Static Pressure Drop	138
Velocity Pressure to Air Flow Calculations	140
The Driving Forces	141
Closed Door Effect	146
Worst Case Depressurization Test	147
Preparing the Building and Combustion Appliance Zone	150
Doing the Test	151
Setup	152
Effect of House Tightness on Zone Tightness	154

Appendix A: Forms — 157

Appendix B: References — 165

Introduction

A technician should always answer the following questions before finalizing an install or service call.

- Is it measurable?
- Did I measure it?
- Did I document my measurement?
- Is it safe to leave the appliance or combustion system running or in operation?
- Did I properly warn or alert the building occupants or supervisors of hazards?
- Would it be safe for my family or me to live or work in this building?

Anyone installing or servicing equipment should examine their own performance.

This text is intended for use by students, installers, service technicians, and first responders.

When installing new equipment and servicing existing systems, it is important to make sure the systems are operating within the measurable definitions of the equipment manufacturer and local code. Proper installation of combustion systems and thorough inspections can reduce loss of life, illness related to combustion by-products, property damage, liability and help reinforce confidence in consumers.

It is impossible to cover every circumstance in the countless situations found in the field. As instructors, authors and collaborators, this material is presented in an effort to provide information that is accurate and field-tested.

Material contained within this manual has been used to conduct seminars around North America for local HVAC and IAQ technicians, manufacturers, inspectors, fire safety and emergency response technicians, fuel suppliers and others.

This text is a collaborative effort of several authors and contributors.

The authors wish to express their gratitude to the growing number of technicians who test and measure the equipment installed or serviced, and who acknowledge the whole building as a factor in controlling air temperature, air quality, system efficiency and safety.

This text contains three sections:

CARBON MONOXIDE SAFETY

This section is intended to help raise awareness to the generation, distribution, detection, remediation and prevention of carbon monoxide (CO).

COMBUSTION ANALYSIS

This segment is intended to present opportunities for economizing fuel & energy use by encouraging the diagnosis of flue gases from theory to manufacturers suggested measurements.

BUILDING PRESSURE DIAGNOSTICS

This section is intended to provide information on easy to use differential pressure tests that help troubleshooting the following problems; fuel pressure and orifice problems, air balancing, appliance venting, duct and building heat loss issues.

AUTHORS

Bob Dwyer - <u>Director of the Bacharach Inc., Institute of Technical Training</u>

Mr. Dwyer has conducted Carbon Monoxide and Combustion Safety Seminars for thousands of people throughout North America.

Mr. Dwyer has been involved with government and public utility energy programs, combustion and environmental testing, HVAC associations, public education and community response coordination. He chaired the Combustion Safety Training Committee in Montana, promoted by the Public Utility Commission, 1991 through 1994.

Rudy Leatherman - <u>Instructor</u>

Mr. Leatherman has been performance testing combustion systems since 1980.

Mr. Leatherman worked with the Energy Office for more than 15 years. His responsibilities included; combustion testing, and evaluating, installing and troubleshooting a wide variety of commercial, industrial and residential HVAC systems. Mr. Leatherman was also involved in educational development and training for thousands of utility company personnel, mechanical contractors, code officials, home inspectors, energy auditors, maintenance technicians, plant/boiler technicians, fire departments, trade associations and others.

Currently, Mr. Leatherman conducts residential/commercial/industrial combustion testing training with Bacharach's Institute of Technical Training throughout North America. He also works closely with Bacharach's Customer Service Representatives to provide technical assistance for customers with application related concerns.

Bruce Manclark

Mr. Manclark has more than 20 years of experience in residential energy conservation.

Mr. Manclark has coordinated training in air sealing, duct sealing, mobile home and standard home weatherization for Pacific Power's Home Comfort program, Oregon Energy Coordinator's Association, Portland General Electric, and the Eugene Water & Electric Board. Mr. Manclark has also published articles in *Home Energy.*
He has offered consumer lighting workshops, and has served as an instructor for auditor and contractor training in computerized audit, blower door and infrared camera techniques, pressure diagnostics, and air and duct sealing.

Mr. Manclark was recently a beta tester for the Energy Conservatory Automated Performance Testing System for Duct Sealing, and for the Lawrence Berkeley Laboratory Aerosol Duct Sealer. He continues to manage the Energy Outlet, an ongoing "market transformation" project, and Delta-T, an energy service company with utility contracts in Oregon.

Ken Kimball - <u>Regional Manager, Technical Trainer for Bacharach</u>

Ken Kimball has a life-long building industry background, having grown up in the family window and door manufacturing business. From 1984 through 1990, Ken owned and operated his own general construction company. Ken has focused his efforts working in the gas heating industry for the past seven years, working with Western South Dakota Community Action Agency, U.S.D.O.E., and A.Y. McDonald Supply. He is a certified trainer and technical advisor in the HVAC and energy industries.

Erik Rasmussen - <u>Bacharach Technical Education Specialist</u>

Erik Rasmussen has a long history and extensive background in the HVAC industry, having been an independent **service contractor and manager for over 12 years. He has completed training and certification in the HVAC** fields of oil burners, gas burners, air conditioning concepts, duct design, heat gain and loss calculations. Erik has been trained by TSSA (Technical Standards and Safety Authority) as an oil and gas burner upgrade certification instructor in the province of Ontario, and has been employed by Bacharach for 5 years, during which time he has conducted combustion and carbon monoxide training and certification programs all across North America.

The main body of text in this book is mostly derived from *"Carbon Monoxide Safety, Combustion Analysis & The Driving Forces of Building Pressures"* by the Bacharach Institute of Technical Training .

Located in Western Pennsylvania since 1909, Bacharach, Inc. is a world leader in the design, manufacturing and service of advanced equipment for the measurement and detection of gases and liquids. Bacharach engineers and scientists have spent decades developing instrumentation and procedures that help assure combustion safety and support indoor air quality.

Important Pressure Measurements and Conversions

PSI (Pounds per Square Inch)
WC" (Water Column Inch)
PA (Pascal)

Atmospheric pressure at sea level = 14.7 PSI
1 PSI = 27.7 WC "
PSI (Standard unit of measure for oil burner pump pressures)

1 WC" = .036 PSI

WC" measurement commonly used in gas manifold pressure delivered to appliance burners (Typical natural gas pressure 3.5 WC") (Typical Propane Pressure 11 WC")

External Static Pressure (ESP) in duct work commonly measured in tenths of WC "

Draft pressure measurements in combustion system vents & chimneys commonly measured in hundreds of an inch WC.

1 WC" = 249.0889 Pascal
Pascal is a common building pressure measurement

1 Pascal = .004 WC"

Common Category I Vent Pressures (systems with non-positive exhaust pressures)
-.02 WC" = 05 Pascal
-.04 WC" = 10 Pascal
-.06 WC" = 15 Pascal

Standard atmosphere or standard conditions: Air at sea level at 59F when the atmosphere's pressure is 14.696 psia (29.92 in. Hg).

Carbon Monoxide

Carbon monoxide (CO) is a toxic gas that can occur in homes and buildings when combustion by-products are generated and allowed to disperse. CO is colorless, odorless, tasteless and is an asphyxiant. **CO is deadly.**

At low concentrations, CO can contribute to existing cardio and respiratory illnesses. It can compound pre-existing health conditions and can go undetected as a cause in premature deaths.

How is carbon monoxide formed? How is it measured?

Carbon monoxide is a result of unburned fuel. Fossil fuels require specific ranges of oxygen and temperature to allow for complete combustion.

CO production is commonly associated with insufficient combustion air. However, excess introduction of combustion air (or insufficient fuel supply) can reduce flame temperatures to the point where CO is produced. When any portion of a flame is reduced below 1128° F, CO will be produced. Flame impingement on heat exchanger surfaces can also result in lowering flame temperature and CO production.

If we completely burn all of the fuel and properly exhaust the by-products of combustion, there should not be any measurable CO.

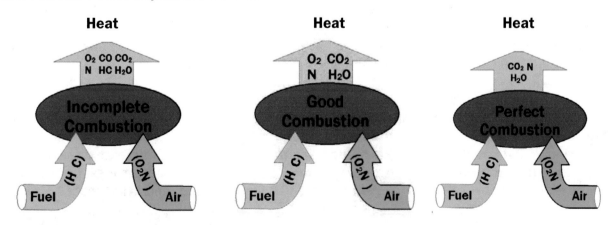

Carbon Monoxide

Certain requirements must be met for combustion to occur. The quality of combustion is dependent upon and rated against the quality of the fuel and its potential to burn completely under perfect conditions.

Fuel that has the potential to burn, like carbon fuels (C), must be surrounded by air or oxygen (O_2) but not flooded with oxygen. Ignition or flash point heat must be enacted and maintained. Fuel, air and heat must all be present or combustion will not occur. In controlled combustion systems, fuel is forced into a combustion zone with limited time constraints because more fuel is being forced into the zone.

TIME - TEMPERATURE - TURBULENCE

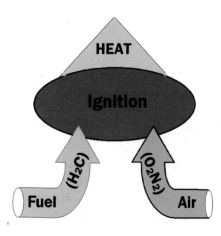

Combustion is a very violent action. **Turbulence** that are controlled help ensure complete burning of fuel and the maintenance of flame **temperature**.

As an example, if the flame temperature is cooled and the turbulence is effected, all the fuel may not have the **time** to burn because it is being forced out of the system by the force of fuel entering. This fragile window of time that fuel has to burn has now been flooded with fuel that spills where it can or follows the strongest drawing forces, out of and below the ignition temperature zone.

The following pictures of gas burners help illustrate how carbon monoxide is formed in some very basic and easy to see ways. It must be remembered that all gas burners are designed to work with controlled fuel mechanisms and in environments that supply sufficient air and oxygen to the fuel at combustion. These burners are designed to burn all the fuel that is sup-

plied.

The ignition temperature of <u>natural gas</u> is between 1100° F and 1200°F with a flame temperature around 3,000°F. The burner with nothing on it has conditions that allow for the complete burning of the fuel, and no CO is produced. When a cold pan of water is set over that flame, a dramatic cooling of the flame occurs and the violent turbulence pushes or spills the unburned fuel out of the flame before it can fully burn, producing CO.

As the pan surface heats up and the water begins to boil, CO generation may cease due to increased heat. (It is recommended that every gas range top burner be tested this way in front of the consumer to help educate them about intermittent carbon monoxide production.) Additionally, it can also be demonstrated that a cooler flame also results in a longer time period for the water to boil. This may be a fraction of fuel savings, but savings none-the-less and obviously less CO generation and dispersion into their living space.

Carbon Monoxide

Carbon monoxide is typically measured in parts per million (PPM); the number of CO molecules present in a million molecules of air.

CO is also measured and referenced as **CO Air Free**. This is a unit of measurement designed to compensate for the excess air to the burner and is only used to express CO levels in a flue gas sample as opposed to ambient air testing.

CO Air Free readings can only be calculated with a test instrument that measures both CO and O_2. Most modern combustion analyzers will make CO air free calculations and provide a CO air free reading in the display.

Depending on the particular piece of test equipment, this CO Air Free reading may be expressed as "CF", "Air Free" or "0% Corrected CO".

This represents the CO levels with no excess air in the sample or with 'perfect' combustion (an unrealistic situation).

An analogy would be comparing two cups of coffee, one of which has had cream added. You still have the same number of coffee 'molecules', but one has been diluted with the cream. A CO Air Free reading would be like performing a calculation on the coffee to remove the cream.

Also be aware that some authorities of jurisdiction require CO readings corrected to a 3% O_2 reading.

Should a test instrument not make the calculations electronically, the CO and O_2 readings can be plugged into the following formula to derive the CO Air Free reading.

$$\frac{20.9}{20.9 - O_2} \times CO = CO\ Air\ Free$$

A furnace with flue gas measurements of 100 PPM CO & 6% O2 would calculate to:
100 X 20.9 = 2090   20.9 - 6 = 14.9
2090 divided by 14.9 = 140.26 PPM CO Air Free

Carbon Monoxide

Another way to determine Air Free readings is to use the following chart. The top horizontal column references the 'diluted' CO reading, the left vertical column references the flue gas O_2 reading.

CO Air Free Reading (PPM)														
CO	25	50	75	100	125	150	175	200	225	250	275	300	325	350
O_2%														
1.0	26	53	79	105	131	158	184	210	236	263	289	315	341	368
2.0	28	55	83	111	138	166	194	221	249	276	304	332	359	387
3.0	29	58	88	117	146	175	204	234	263	292	321	350	379	409
4.0	31	62	93	124	155	186	216	247	278	309	340	371	402	433
5.0	33	66	99	131	164	197	230	263	296	329	361	394	427	460
6.0	35	70	105	140	175	210	245	281	316	351	386	421	456	491
7.0	38	75	113	150	188	226	263	301	338	376	413	451	489	526
8.0	41	81	122	162	203	243	284	324	365	405	446	486	527	567
9.0	44	88	132	176	220	263	307	351	395	439	483	527	571	615
10.0	48	96	144	192	240	288	336	383	431	479	527	575	623	671
11.0	52	106	158	211	264	317	369	422	475	528	581	633	686	739
12.0	59	117	176	235	294	352	411	470	528	587	646	704	763	822
13.0	66	132	198	265	331	397	463	529	595	661	728	794	860	926
14.0	76	151	227	303	379	454	530	606	682	757	833	909	984	1,060
15.0	89	177	266	354	443	531	620	708	797	886	974	1,063	1,151	1,240

For example: A burner operating with a 3.0% O_2 and a 100 PPM 'diluted' CO reading equates to a 117 PPM CO Air Free reading.

Check with local codes to determine the allowable flue gas CO levels in your community. Should local codes not address this issue, national authorities of jurisdiction such as the American Gas Association (AGA) require flue gas CO levels be under 400 PPM Air Free.

Chapter 15 in the Uniform Mechanical Code 2000 describes flue gas CO measurement as "not being greater than 0.04 percent." Chapter 15 is new to the code and this specific measurement has originated from this CO Air Free calculation.

However, years of test experiences have found that CO levels in excess of 50 PPM Air Free (once the burner has reached steady state conditions) are indicative of a combustion problem and that CO Air Free levels in excess of 400 PPM Air Free upon ignition represents a 'rough' ignition.

Signs and Symptoms of Carbon Monoxide Poisoning

- Confusion
- Dizziness or headache
- Eye and upper respiratory irritation
- Fatigue
- Wheezing or bronchial constriction
- Persistent cough
- Increased frequency of angina in persons with coronary heart disease
- Elevated blood carboxyhemoglobin levels

If carbon monoxide is measured within the living or working space of a building, the following diagnostic approach may aid in the discovery of its source and perhaps its affect on the inhabitants.

This discovery begins with questions.

If a caller reports a carbon monoxide alarm, the dispatcher/technician must find out if anyone at the location has any of the above symptoms and may choose to further explore the symptoms through the following questionnaire. In some jurisdictions, a carbon monoxide alarm reported means immediate dispatching of a response team regardless of the symptoms.

- When did the symptoms or complaints begin?
- Does this symptom or complaint exist all the time or does it come and go?
- Is the symptom associated with a particular location or time of day?
- Is the symptom seasonal in nature?
- Does the problem seem to improve, your health improve after you leave a specific place?
- Are the symptoms associated with a change of workplace or living locations?
- Is anyone else in your house or building have similar symptoms or complaints?
- Is the symptom associated with the use of any heating or cooking equipment?
- Do you have an attached garage? Do you warm your car up inside with door open?
- Are you a smoker or around smokers during the times of your discomfort?
- Is charcoal being burned indoors in a grill, fireplace or other cooking device?
- Is there an odor present when heating, cooking or other combustion appliance in use?
- What types of combustion equipment are in use?
- When was the last time the combustion equipment was serviced?
- Does any of the combustion equipment seem to be in disrepair?

Regardless of our role in the industry, we have opportunities to discover the source or sources of carbon monoxide. **However, we have to look at the house or the building as a system.**

Carbon Monoxide

Health Effects of Carbon Monoxide Poisoning

As professionals in the field we come into many households and commercial buildings containing people with medical problems. This includes households with elderly or anyone with heart disease, pregnant women, infants, children and adults with asthma, or people of any age experiencing temporary poor health like the flu, colds, seasonal allergies and others.

We have the opportunity to diagnose and eliminate this poison which may be contributing to poor health conditions. Let's examine how carbon monoxide poisons us.

When carbon monoxide is inhaled into the lungs it bonds with hemoglobin in blood, which forms *Carboxyhemoglobin (COHb)*. This condition displaces oxygen in the blood stream and effects all major organs and muscles.

It has been determined that carbon monoxide molecules bond with hemoglobin in blood over 200 times more easily than oxygen molecules. Suffocation occurs from the inside out.

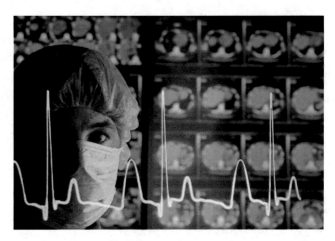

The health effects of CO depend upon the concentration in the air and the duration of the exposure. Extended exposure to high concentrations will lead to unconsciousness, brain damage or death. The very young, elderly and those with certain pre-existing health conditions are more vulnerable at lower concentrations of exposure for longer periods of time, and may suffer similar effects as those exposed to higher concentrations for brief periods.

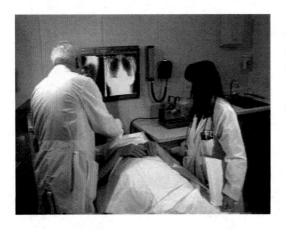

Healthy adults may not show severe effects to low concentrations of carbon monoxide. Headaches, constant stuffiness or head pressure are common symptoms of early CO poisoning and may be the prelude to a worsening condition. Sometimes, these conditions can go undiagnosed.

Respiratory problems, chronic heart disease, dizziness, vomiting, confusion, general weakness of the body or flu-like symptoms may be a result of CO poisoning.

Recent medical studies attempting to further understand low level CO poisoning have found that blood vessels are a major site of damage in the brain; especially the cells that line the inner wall of the vessels called the endothelium. This damage occurs relatively early during exposure to CO. Additional studies suggest this could be happening with lower concentrations of CO over longer periods of time.

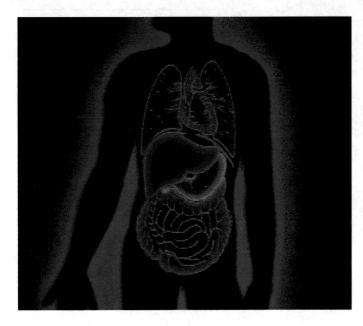

These recent studies also suggest that CO molecules bond themselves to the same proteins in human blood as Nitric Oxide (NO). Nitric Oxide is a naturally occurring vasodilator in the body. It tells the body to make blood vessels relax and widen. Nitric oxide functions as a gaseous signaling molecule. Signaling molecules in the nervous system are called neurotransmitters.

An excess of Nitric Oxide creates an imbalance in the bloodstream and is harmful to brain cells and other tissues. This imbalance makes Nitric Oxide available for biochemical reactions that would not normally occur within the cell, These reactions produce tissue-damaging oxidants and free radicals.

Research shows that exposure to increased concentrations of CO result in more Nitric Oxide being released by cells, resulting in cell death. Lower doses of CO result in less cell death.

Hopefully, the end result of CO and human impact studies will result in improved general understanding of CO exposure and a more aggressive and preventative treatment.

In the field, simple observations or findings may alert you to a potentially dangerous condition. As an example, the occurrence of illness in household pets concurrent with or just preceding the onset of a patient's own illness is an alert to the possibility of CO poisoning.

Due to their smaller size and generally higher metabolic rate, pets may be more obviously and severely affected by CO intoxication than their owners.

Arterial blood sampling has been demonstrated to be the most traditional and accurate way of determining COHb %. This procedure is painful, expensive and not readily available for field use. General practitioners and others in the health profession do not routinely draw blood samples from every patient showing symptoms that can now be associated to CO poisoning.

Carbon Monoxide

The most common misdiagnosis of CO poisoning is a "flu-like" syndrome. Additional misdiagnosis include food poisoning, depression, coronary artery disease, arrhythmia and functional illnesses among others. Blood sampling for CO in the field is not practical. However, breath analysis for CO with a Bacharach BAM unit is.

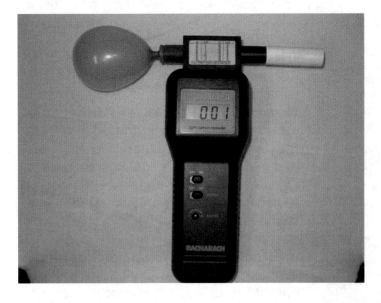

This Breath Analyzer Module (BAM) easily attaches to the top of the Monoxor® II.

The sanitary and disposable mouthpiece and balloon accessories attach to the BAM.

The victim inhales, holds breath for 20 seconds, exhales about ½ of that breath to purge the mouth and throat before inflating the balloon to about 5 inches in diameter.

In the graph, the bottom horizontal column is CO in PPM and the registered PPM from the instrument is noted. The vertical column is the approximate COHb %. Read up from the highest recorded CO PPM until it intersects the diagonal line. Read to the left for COHb%.

While a University of Pittsburgh study validated the accuracy of determining COHb levels in the blood utilizing the Breath Analysis Module, several professionals in the field were uncomfortable with this direct correlation.

Further research is being conducted. The bottom line is that when testing a healthy non-smoker, if you find a significant CO readout in the display, the individual has undoubtedly been exposed.

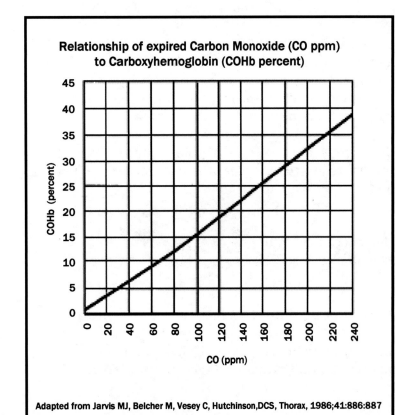

Adapted from Jarvis MJ, Belcher M, Vesey C, Hutchinson,DCS, Thorax, 1986;41:886:887

Carbon Monoxide

The main therapy for CO poisoning is administering supplemental oxygen and ventilatory support and the monitoring of heart rate. The goal of oxygen therapy is to increase the O_2 level of the blood. O_2 therapy and observations should continue long enough to prevent additional poisoning once carboxyhemoglobin unloads from the cell.

No set guideline for length of therapy is given and health professionals readily admit that there is a tremendous amount of unknown health effects from CO exposure.

Historically Accepted Medical Symptoms of Carbon Monoxide Poisoning

Slight headaches, tiredness, dizziness, and nausea after 2-3 hours	200 PPM
Frontal headaches within 1-2 hours, life threatening after 3 hours	400 PPM
Dizziness, nausea and convulsions within 45 minutes. Unconsciousness within 2 hours. Death within 2-3 hours.	800 PPM
Headache, dizziness and nausea within 20 minutes. Death within 1 hour.	1,600 PPM
Headache, dizziness and nausea within 5-10 minutes. Death within 30 minutes.	3,200 PPM
Headache, dizziness and nausea within 1-2 minutes. Death within 10-15 minutes	6,400 PPM

Keep in mind that these historically accepted concentrations originated from an early study conducted by the Bureau of Mines in 1925 and again in the late 1960's on US Military volunteers. These studies involved healthy young adult men.

When we find carbon monoxide inside the buildings we are servicing that exceeds the levels outside, our role becomes more vital. **Who is in the building?** Are they all young, and healthy adults? Your inquiries about the general health of the inhabitants may reveal minor or compounded illness symptoms associated to the levels you measured.

ALL EFFECTS CAN VARY SIGNIFICANTLY BASED UPON AGE, SEX, WEIGHT AND OVERALL STATE OF HEALTH.

It is vital that a cooperative relationship guides all activity associated with carbon monoxide incidences. Just as consistent step-by-step procedures be used when in a home or building, a diagnostic checklist concerning poor health patterns or symptoms should be used. This checklist may be useful to everyone in the field as well as health care professionals and emergency responders. **Carbon monoxide is everybody's business.**

Carbon Monoxide

The following information is a synopsis of a study on the health effects of CO conducted in the United Kingdom.

Please note that because this study was conducted in the United Kingdom spelling of medical terms may vary.

CARBON MONOXIDE
Invisible Destroyer of Health and Life

Carbon Monoxide

Introduction and Key Findings

Carbon monoxide (CO) can seriously damage your health. This survey of 77 individuals poisoned by CO over several years has revealed health problems which are both debilitating and persistent.

The survey was undertaken by CO Support to:

- raise awareness of the symptoms of chronic exposure to CO;
- identify the health consequences of such long term exposure, and
- identify those who may be at risk.

The key findings of the study are:

1. Those suffering from chronic exposure to CO experienced a wide range of symptoms, including memory loss, severe muscular pain, headaches, tiredness and dizziness.;
2. In many cases, these symptoms continued for years after the exposure ceased. Although some people have recovered completely, a significant proportion remain permanently incapacitated and unable to work;
3. GPs failed to diagnose chronic exposure to CO. In only one case out of the 77 studied was exposure identified on the basis of symptoms alone;
4. Misdiagnoses included flu, viral infection, depression, ME, and psychosomatic illness. Often no diagnosis was given at all;
5. In the majority of cases, the presence of CO was discovered by servicing or investigation of the offending appliance. In some cases warning was given by an alarm or detector. In others, the collapse of one family member drew attention to the problem;
6. In many cases, regular servicing of the appliance failed to identify the problem. In some cases servicing took place regularly during an exposure lasting several years;
7. Around 70% of chronic exposures took place in people's own homes. These were usually houses rather than flats, and many were owner occupied;
8. Two thirds of sufferers were women, with most aged between 30 and 45 years;
9. Very few sufferers were offered a carboxy haemoglobin (COHb) test to determine the extent of their exposure. Where tests were performed there was also evidence of misinterpretation of the results by hospitals and GPs.

Carbon Monoxide

The ways in which CO damages the body are many and diverse. However medical research into its effects has so far been limited, with the most recent medical article to explore chronic exposure to CO being published in 1936. It is our experience that quite low levels of CO exposure, experienced over a prolonged period can cause lasting symptoms.

CO exposure can cause tiredness, headaches, depression, dizziness, nausea, breathlessness, flu-like symptoms and muscle pains, as well as numerous other effects. However, all of these symptoms are also associated with other diseases, such as flu, ME or asthma. As a result, there is widespread misdiagnosis by doctors and chronic exposure to CO remains a vastly under-recognised problem.

CO Support
The founder of CO Support had her health destroyed by a blocked flue to the gas fire in the living room. Her response was to set up a support group for other sufferers, and to gain as much information as possible about the long term effects of CO on the human body. Even with limited publicity and resources, the number of people seeking help since then has been large, and the membership of CO Support has been growing rapidly.

In the absence of medical evidence on the effect of long term exposure to CO[1,] the members of CO Support decided that they should explore the problem scientifically. A professionally-designed questionnaire was constructed which probed the cases that had come to light. After excluding those where the evidence was insufficient or inadequately described, 77 cases, in a total of 49 different incidents, have been explored in depth. This paper tells their story; a more detailed technical paper has also been prepared which is available upon request.

Florence Nightingale could be one of the early recorded victims of CO poisoning. During her work in the Crimea, she sat up wrestling with her statements for the War Office night after night. It was bitterly cold, the stove sent out from England would not draw, and the charcoal brazier used to heat her room made her head ache. "They are killing me" she wrote in letters home to her Aunt May[2].

Her symptoms included headache, nausea, chronic fatigue leading to collapse, ear-ache, laryngitis and deep muscle pain in her back. These are all classic symptoms of CO poisoning. At the time doctors diagnosed congestion of the spine.

After her return to England she experienced continuing ill health, and unable to walk any distance she held court with Government officials from her sofa. At one point Miss Nightingale became bed-ridden and did not leave her room for six years.

The nurse who did so much to improve health care for others needed help herself from a relatively young age, quite possibly as a result of fumes being emitted from burning charcoal briquettes.

Carbon Monoxide

The Study

The 77 cases examined involved 12 where those exposed became unconscious ("unconscious cases") and 65 where the exposure caused severe symptoms but fell short of causing unconsciousness (termed "chronic cases"). The distinction is important because many doctors wrongly believe that, provided you do not become unconscious, the effects of exposure disappear automatically and completely.

In order to ensure that the symptoms are genuinely different from those expected of people of the same age and background, a carefully-matched control sample was also established. This control group formed a benchmark against which the health of the exposed group could be assessed.

The Results

Profile of Sufferers

Some two thirds of those who suffered chronic exposure to CO were women. The most commonly affected age groups were 30 to 45. This profile suggests that women who spend much of their time in the home, possibly looking after young families, may be one of the groups particularly at risk.

Table 1: Profile of Groups Exposed

	Chronic Group	Unconscious Group
% Female	63%	67%
% Male	37%	33%
Average age	36	48
Most common age	35	-
Age range	2 TO 76	27 TO 81

In the majority of cases all members of the household were affected. However, in nearly 40% of cases, one or more members of the household were not noticeably ill, probably as a result of lower usage of the rooms served by the offending appliance.

The average length of exposure was four years for the chronically exposed group, with the typical exposure being 2 to 3 years. The length of exposure for the unconscious group was shorter, at 2.5 years on average. The time since ceased varied from a few months to 22 years, with an average of nearly 4 years.

The cases that contacted CO Support and were included in the study came from all over Great Britain.

As shown by Figure 1, around 70% of the incidents occurred in houses, with under 20% occurring in flats. Of these houses, a significant proportion were owner occupied (40%). Fireplaces were responsible for around 60% of incidents, and central heating for one third. Just over 10% occurred in the workplace.

Carbon Monoxide

Figure 1: Type of Property

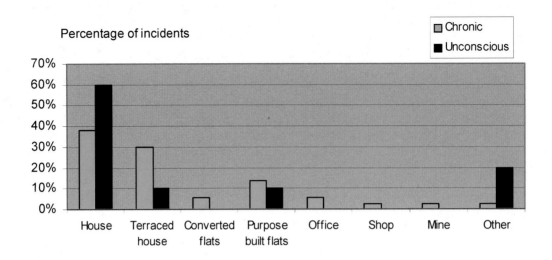

Discovery

In less than 10% of cases was CO exposure diagnosed by the doctor, and in only one case was the diagnosis on the basis of symptoms alone. In the majority of cases, the presence of CO was identified through servicing or investigation of the offending appliance. CO alarms are increasingly being installed, as awareness of the problem has grown, and these alerted sufferers in 8% of the incidents.

Figure 2: Method of Discovery

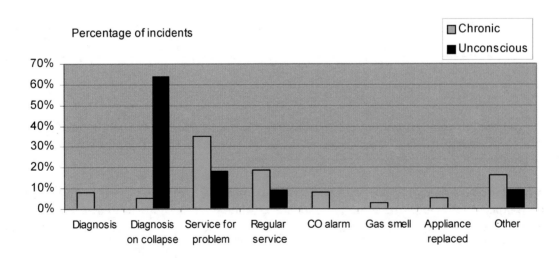

Another surprising finding is that relatively few of the incidents (under 10%) were found through regular servicing of the appliance, even though over 50% of the households concerned had their appliances serviced regularly. In none of the unconscious cases was the fault discovered through the regular service. This clearly demonstrates a need for more stringent enforcement of standards for the installation and servicing of appliances.

Symptoms

CO poisoning creates a wide variety of effects, many of which are non-specific to CO. The study showed that;

- On average, symptoms such as headaches and dizziness were experienced for nearly two years prior to CO being identified as the problem;
- In many cases, the symptoms persisted for two or three years after exposure ceased, and are often still continuing.
- During the exposure, the four most common symptoms (experienced by over 80% of those exposed) were tiredness, headaches, pains and cramps, and nausea/sickness.

Figure 3: Symptoms during and after exposure to CO

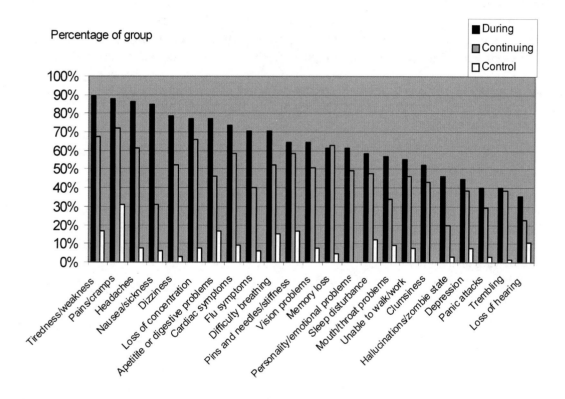

Figure 3 shows the wide range of symptoms experienced by the chronic group during and after exposure to CO, which contributed to the difficulty of diagnosis.

- In the chronic group symptoms included lack of concentration, cardiac problems, flu-type symptoms, difficulty breathing, cramps, memory problems, personality changes, clumsiness and depression.
- In the control group, the incidence of problems was dramatically lower;
- The experience of the unconscious group was similar, although the proportion suffering from each symptom during the period of exposure tended to be slightly higher.

Carbon Monoxide

Figure 3 shows also that for many sufferers the symptoms have persisted long after CO was identified and removed.

- Only four of the symptoms, nausea and sickness, hallucinations, flu symptoms and digestive problems showed significant improvements;
- In general there was only a relatively slight improvement in the incidence of symptoms after CO was removed; and indeed
- The number of respondents identifying memory loss as a problem increased after exposure ceased.

The study found that many people suffered pain following exposure to CO:

- During exposure 88% of the chronic group suffered from flu like aches, chest pains, pains in the arms and legs and/or neck and back pain;
- Neck and back pain and deep muscle pain have remained the most persistent problems;
- The incidence of pain amongst the unconscious group was even greater than for the chronically exposed group, with over 90% suffering from such problems during exposure;
- Over 70% of both groups continue to suffer from some degree of pain, which represents only a small improvement in the percentage of those suffering since the exposure ceased (16%).

Other effects reported by sufferers included greater sensitivity to chemicals such as chlorine or bleaches, paints and car fumes. Well over half the sample also found that their tolerance of alcohol had deteriorated since exposure.

Ability to work

- Over 45% of the members of the chronic group were still unable to work at the time of the survey, an average of four years after exposure ceased.
- 75% of those rendered unconscious were still unable to work two and a half years after the source of CO was removed; and
- As a result of CO exposure, household income halved in the unconscious group and fell by over 20% in the chronic group.

Medical Diagnosis

Partly because of the wide variety of symptoms, and partly because other common illnesses exhibit similar symptoms, the study showed that GPs very rarely diagnosed chronic exposure to CO on the basis of the symptoms alone.

Nearly 60% of the chronic group and 75% of the unconscious group were given no diagnosis for the symptoms they experienced during exposure;
The most common misdiagnosis (given to around 15% of sufferers) was that they were suffering from a flu or a virus;
8% of the chronic group were diagnosed as suffering from depression, and 5% from ME; and
Nearly 10% of the unconscious group were told that they had a psychological problem or were suffering from a psychosomatic illness.

Since the identification of CO exposure, patients have been treated for a wide range of illnesses. These include depression, ME, heart problems/angina, asthma, Parkinsonism, diabetes, arthritis, MS, fibromyalgia, crohnes disease, and anaemia.

There were also a number of indicators which suggest that there may in fact be considerable misdiagnosis of chronic exposure to CO as ME or Chronic Fatigue Syndrome. Three people within the chronic group were misdiagnosed as having these conditions, while many of the CO sufferers in our study have experienced, and continue to experience, debilitating tiredness with the muscle pain which is thought to be a characteristic of ME[3]. Also, the age/sex profile of our chronic group is very similar to what has come to be recognised as the profile of a "typical" ME patient[4].

The incidence of misdiagnosis could be greatly reduced with a simple blood or breath test, taken by GPs to check for the presence of CO. However as shown by Figure 4, this test was rarely offered to those suffering from long term CO exposure.

Figure 4: COHb Test

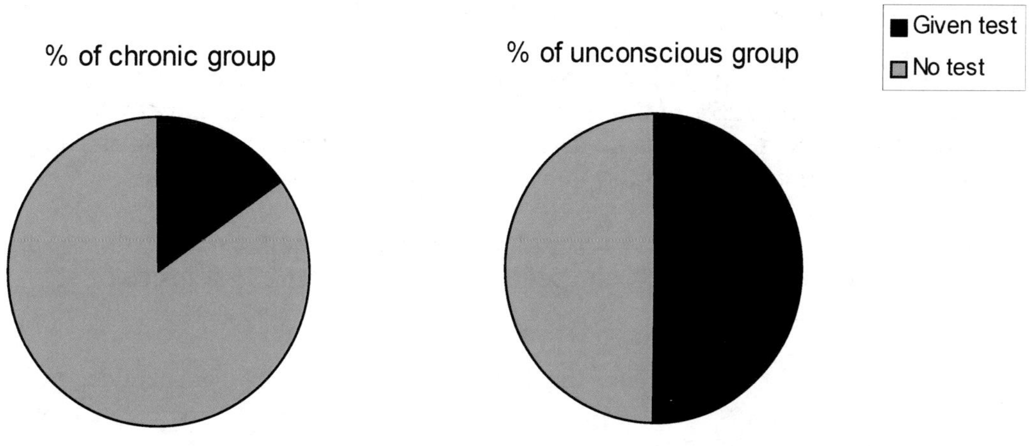

Carbon Monoxide

Medical Support

The most common misdiagnosis of CO exposure was no diagnosis at all. As Figure 5 shows, in many cases, GP's were felt to be unsupportive.

Figure 5: Doctor's reactions

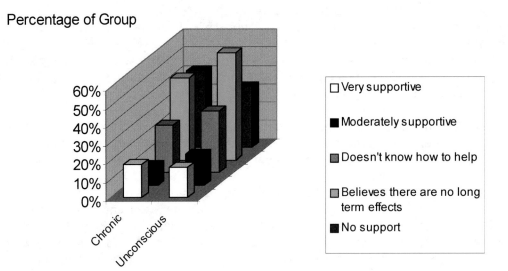

- Only 29% of the chronic group and 34% of the unconscious group felt their doctors were supportive;
- A substantial proportion felt their doctors did not know how to help (26% of the chronic group and 33% of the unconscious group);
- Many doctors believed that exposure to CO has no long term effects: 45% of the chronic group and 58% of the unconscious group have been told this by their doctors.

These findings reflect the relative lack of training in toxicology given to doctors. A total of 1 to 12 hours is devoted to toxicology, of which Carbon Monoxide is just one part. The amount of training given in this area is an issue that needs to be addressed by the medical profession.

The seriousness of the health problems can be judged by the numbers referred to consultants after the discovery of CO. Around 25% of the chronic group had consulted cardiovascular/respiratory specialists and/or neurologists. Other specialists consulted included psychologists, lung disease specialists and rhuematologists. Notably, only 5% had seen a toxicologist.

The proportion of the unconscious group consulting neurologists and toxicologists was significantly higher than for the chronic group; 42% had seen a neurologist and 25% a toxicologist.

Hyperbaric oxygen treatment (receiving oxygen under pressurised conditions) can aid recovery from CO poisoning, and in particular help to avoid reperfusion brain damage. However, such facilities are in short supply, and most hyperbaric units have a policy of helping only those who were found unconscious. Our study found:

- Only 18% of unconscious cases and less than 2% of chronic cases were offered such treatment.

CO exposure can be established by a relatively simple carboxy haemoglobin (COHb) blood test. However as noted earlier:

- Only 15% of the chronic sample were given such a test; and
- Even when undertaken, the test was usually delayed for hours or days.

This renders the interpretation of the results suspect due to the rapid elimination of COHb from the blood once the patient is removed from the source of CO. (COHb has a half life of 4.5 to 5.5 hours[5].) Today, relatively cheap hand-held monitors are available to provide accurate measurement of the level of CO in the bloodstream, but very few GP surgeries either know about or possess such equipment.

Wider evidence - failure of diagnosis

Many of the findings of the study are supported by other research and evidence. For example, CO Support's finding of widespread diagnostic failure by GPs is supported by a recent UK study.[6] 200 GPs were given a description of the symptoms of carbon monoxide poisoning, namely nausea, headache, lethargy and flu-like symptoms, and asked for possible diagnoses. Not one doctor raised CO as a possibility.

The potential for misdiagnosis of CO exposure has also been highlighted in the medical literature[7]. Moreover, a 1989 study suggested that there was extensive under-recognition of the number of deaths due to CO in England and Wales for the year 1985[8]. By collating detailed hospital records, this study found a total of 1,365 deaths that were attributable to carbon monoxide poisoning. For that same year, British Gas reported 75 deaths from burnt gas poisonings in its annual report, and official statistics cited 475 hospital admissions and only 10 deaths as due to CO poisoning.

Clearly the available figures relating to both exposures and deaths show such disparity as to render them unusable. A central collation should be arranged to ensure that we have a full and clear picture of the true extent of exposure to CO.

In sum these findings suggest that exposure to CO remains a largely hidden problem. Indeed, the sample on which the present study is based involved only those who had contacted a small and relatively unknown charity for help. The fact that over 100 such people emerged during a period of under one year is suggestive that the problem may be more widespread than is commonly acknowledged.

Carbon Monoxide

Possible prevalence of Chronic CO Poisoning

Wider evidence also suggests that CO may be an unrecognised problem. As there has been no systematic investigation, the available evidence is incomplete and largely circumstantial. However, the following facts highlight the changes which have occurred in the circumstances of householders and tenants, which show cause for concern:

- Central heating is now installed in a majority of homes; in 1965 only 7% of homes had central heating, compared with 75% in 1993.
- Gas central heating has risen from 24% of households in 1976 to over 60% in 1991.
- Nearly half of all homes now have full or partial double-glazing, compared with only 3.9% in 1970. It is possible that this has led to reduced external ventilation of rooms and appliances with a consequent build up of fumes.

In addition, there are a number of epidemiological puzzles that have emerged over a similar period:

- Britain appears to have much larger numbers of 'excess winter deaths' than other countries – perhaps twice the level of Scandinavia or Germany, for example[9]. Hidden influenza was identified as potentially a significant factor that should be investigated further.
- A recent study[10] highlights an urban-rural gap in death rates, which requires investigation. The worst places appear to be those cities with higher proportions of rental accommodation.
- There is evidence of a rise in dizziness, giddiness, headache, chest pains and nausea of between 30 and 40% since 1970[11].
- The steep rise in the incidence of asthma is well known. Although there is no known direct link to CO exposure, it has been suggested[12] that a factor contributing to the increase might be increased sensitivity to allergens. Exposure to CO may also be associated with exposure to nitrogen dioxide, which has been cited as a possible cause of respiratory problems. One study found an association between gas appliances (particularly cookers) and respiratory problems[13]. The rise in asthma has occurred particularly among the lower social classes, females and children and old people. Between 1971 and 1992, asthma incidence rose by 114% for males and 165% for females.
- Recently diagnoses such as ME and Chronic Fatigue Syndrome have become more common. The similarities between the symptoms of these conditions and those of chronic CO exposure raises the possibility of misdiagnosis[14].
- A study of Aberdeen schoolchildren[15] found that between 1964 and 1994 the incidence of wheezing and shortness of breath had doubled.
- Between 1971 and 1992 the incidence of migraine has increased by 23% for males and 50% for females[16].
- A recent major survey of GP's shows that 18.3% of patients were substantially 'fatigued' for six months or longer, and that 30% of these patients combined fatigue with muscle pain[17].

The general picture is one of widespread rises in the potential for domestic exposure to CO accompanied by unexplained increases the symptoms of chronic exposure (such as headaches, dizziness, respiratory problems and heart failure), together with increases in conditions which might reflect misdiagnosis (such as ME and influenza).

Taken together, these factors reinforce the urgency of further investigation into the extent and consequences of chronic exposure to CO.

Conclusions

This study refutes a number of widespread myths about the circumstances and effects of long term sub-lethal exposure to carbon monoxide.

MYTH:	Only students and holiday makers in poorly maintained flats or apartments are at risk from CO.
FACT:	Everyone is at risk. Long term exposure to CO typically occurs in people's own homes. It is not only rented properties either: exposure can and does occur in owner occupied homes, and across all income groups.
MYTH:	Chronic CO poisoning is rare.
FACT:	Chronic exposure to CO is grossly undiagnosed. Symptoms are often confused with flu, viruses and according to our survey, ME.
MYTH:	CO either "kills you or it doesn't", and if it doesn't you will not suffer any long term effects.
FACT:	CO poisoning can have a wide range of debilitating or permanent effects. These include memory loss, constant pain and unending tiredness. Many sufferers remain incapacitated and unable to work.
MYTH:	Getting your appliances serviced regularly will keep you safe.
FACT:	It may, but there are no guarantees. Too often servicing is perfunctory and fails to identify the real problems, especially in the flue.
MYTH:	Everyone in the household will be affected by CO, including the pets.
FACT:	One or more members of the household may remain unaffected, depending upon which rooms give the greatest exposure.

Carbon Monoxide

Carbon monoxide is an invisible, insidious problem, which remains hidden by a vicious cycle of misdiagnosis. Chronic exposure to CO is regarded as rare (and having no permanent effect). Therefore doctors are not aware of its importance and misdiagnose CO poisoning. If most cases of poisoning are missed it is no surprise that chronic poisoning is said to be a rare occurrence.

Figure 6: Cycle of Misdiagnosis

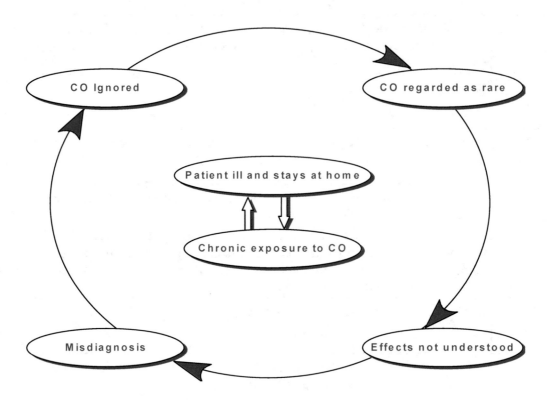

There is an urgent need for:

- a concerted public awareness campaign to highlight the dangers and symptoms of <u>chronic</u> exposure to CO;
- improved COHb testing. Newly available non-invasive breath tests should be carried out routinely by hospitals and ambulances on those suffering from the symptoms of chronic CO exposure;
- raised awareness amongst the medical profession of the symptoms, testing and treatment of chronic CO exposure;
- research into the long term effects of chronic exposure, and a cure sought for the consequent tiredness and pain;
- a series of trials at GP surgeries to establish the true extent of long term sub-lethal exposure to CO. (CO Support are planning a pilot study this winter using breath analysers);
- comprehensive and reliable collection of statistics relating to CO exposures and deaths; and
- improved training and enforcement of standards in the installation and servicing of appliances.

Footnotes

1. CO Support has since heard of two research projects which have recently started in America. It will be a year or two before these results become available.

2. Woodham Smith,C "Florence Nightingale", Richard Clay and Co, 1950

3. Pawlikowska, Chalder et al, "Population based study of fatigue and psychological distress", BMJ, volume 308 19 March 1994.

4. Wilson, Hickie et al, "Longitudinal study of outcome of chronic fatigue syndrome", BMJ, volume 308, 19 March 1994

5. KK Jain, "Carbon Monoxide Poisoning", Warren H Green, 1990, p31

6. Henry, J. Study conducted by NOP, 1994

7. Lowe-Ponsford, F and Henry, JA "Clinical Aspects of Carbon Monoxide Poisoning" in Adverse Drug Reactions Acute Poisoning Review, 1989 8(4) p217-240

8. Meredith, T and Vale, A. "Carbon Monoxide Poisoning", BMJ Vol 2 1989.

9. Curwen, M, "Excess winter mortality: a British phenomenon?", Health Trends, Vol 22 No. 4, 1990/91, pp169-175.

10. Dorling, D, Report for the Joseph Rowntree Foundation, 1997.

11. "Morbidity Statistics from General Practice", Second, Third and Fourth National Studies, OPCS and Department of Health. The figures cover the period 1970-1982; because of a change in the classification regime for diseases, a clear trend to the current date cannot be established.

12. DOH Committee report

13. Jarvis D, Chinn S, Luczynska C and Burney, P. "Association of respiratory symptoms and lung function in young adults with use of domestic gas appliances" The Lancet Vol 347, 1996 p426 to 431.

14. A recent study suggests that a majority of chronic fatigue syndrome patients improved over a three year period, the same sort of timing as for CO exposure. Wilson, Hickie et al, "Longitudinal study of outcome of chronic fatigue syndrome", BMJ, volume 308, 19 March 1994.

15. Ninan, T and Russell, G, "Respiratory symptoms and atopy in Aberdeen schoolchildren: evidence from two surveys 25 years apart", BMJ volume 304, April 1992, pp873 to875.

16. "Morbidity Statistics from General Practice", Fourth National Study 1991-1992, OPCS and Department of Health, 1995, p72.

17. Pawlikowska, Chalder et al, "Population based study of fatigue and psychological distress", BMJ, volume 308 19 March 1994, pp763 to766.

Sources of Carbon Monoxide Poisoning

Auto exhaust is thought to be **the number # 1** cause of accidental CO poisoning in North America and has been reported to be the cause of around **60%** of carbon monoxide alarm responses.

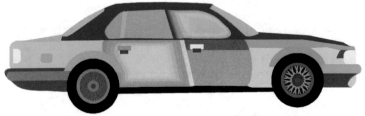

Notice how many people let their automobiles warm up inside garages with the door open and for how long before they back out and close the door with their automatic door control. Typically, any gasoline engine produces the highest CO levels during a cold start.

CO gets trapped inside the garage and can easily disperse into the rest of the building through unseen but loose fitting construction connections (like wiring penetrations, framing joints, ductwork seams, door jambs and other areas).

Building pressure and temperature variations work as siphon points and air exchange locations.

Carbon monoxide in homes does not always come from traditional sources. The service and HVAC industry has been targeting cracked heat exchangers as the leading source of CO poisoning (almost exclusively without test instrument verification until about 1985). Recently, more thorough testing for CO suggests that unvented, poorly installed, un-maintained and misused gas and oil appliances are **the 2nd** leading cause of CO alarm response, and may constitute as much as **20%** of CO alarm call sources.

The 3rd leading cause of CO exposure appears to be due to vented atmospheric, natural drafting appliances which backdraft into the structure and may account for

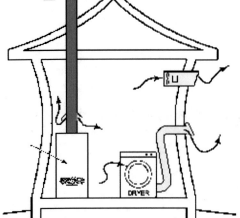

19% of the CO alarms going off.

Intermittent backdrafting of CO laden flue gases complicates source investigations. Improperly sized and installed vent systems, old vents in need of repair, and competing building pressures contribute to this affect.

Even if no significant levels of CO are measured in the flue gases, keep your eyes open to indications that spillage is occurring during other periods of time. Soot or rust on draft hood surfaces or above the burner areas are common examples.

It should be noted that cracked heat exchangers make up one of the smallest percentages of CO alarm response causes (less than 1%).

Carbon Monoxide

Furnaces with cracked heat exchangers that are frequently condemned by technicians or utility personnel are often red tagged and shut off, only to be returned to operation by a homeowner.

A furnace left in operation while a known crack exists represents a tremendous health, safety and liability concern and should be thoroughly tested for CO production. Test results can then be presented to the homeowner emphasizing a very real and present danger.

If You Don't Test, You Don't Know!
Warn and Inform!

THERE ARE MANY SOURCES OF CO!
HOW MUCH ORIGINATES OUTSIDE?
HOW MUCH ORIGINATES INSIDE?

ALWAYS ENTER BUILDINGS WITH TEST INSTRUMENTS ON!

Monoxide

Understand How Your Test Instruments Calibrate

While working around potential CO generation sources it is important to be certain that the work site is safe for the technician. Manufacturers of CO test instruments typically design the instrument to calibrate either 'manually' or 'automatically'.

Automatic calibration results in the instrument calibrating to 0 regardless of ambient CO levels. Bacharach's PCA and Model 300 perform calibration in this manner. To provide a margin of safety, these instruments will display a CO sensor error should ambient CO levels be in excess of approximately 50 PPM during the 60 second calibration cycle.

These instruments can still be used for CO source investigation and protection of the technician; it simply has to be calibrated outside, before entering a building.

Bacharach's Monoxor II is specifically designed for CO source investigations. It performs a 'manual' calibration during its brief warm up period. If there are 25 PPM's (CO) in the ambient air, for example, the instrument immediately displays that reading.

During development of the newest combustion analyzer (the **Fyrite Pro**), it was decided that the type of calibration should be up to the technician. This instrument can be easily set up for either 'manual' or 'automatic' calibration.

For instructions on instrument calibrations:
www.bacharach-training.com

Carbon Monoxide

CO Alarms are Warning Devices

How many structures do we enter that have working and **properly placed carbon monoxide alarms or detectors?** Do we sell or recommend the placement of these alarms?

Before we tell the consumer where to put CO alarms, we need to understand how air movement in buildings and duct systems impact the safe and efficient heating and cooling operation.

Also, we need to determine how combustion, draft, make-up/combustion air and other factors effect the building's operation as a 'system'.

Is there a best location for CO alarms? Do we understand how carbon monoxide alarms work?

Do we know the alarm standards? Can we explain these to the consumer?

Are we **alerting the consumer to potential and foreseeable hazards** associated with any of their combustion systems? Are we educating the public as we service them or are we merely billing them? We have opportunities to educate them as well as generate more business.

<u>**Every building we enter that has external influences on the combustion system should have a CO alarm.**</u>

If one is not in place upon your visit, suggest their installation. This should also include all electric homes with attached garages.

It is always suggested to recommend the installation of a home CO alarm. It is not uncommon to find heating contractors and others, after having notified **the Authority Having Jurisdiction** about a cracked heat exchanger, to leave "loaner" CO alarms and the furnace in operation until repair or replacement can occur. Sometimes the consumer purchases the CO alarm or the contractor builds that into the replacement bid as an option. They may even give one as a 'free' service.

A CO alarm may provide additional protection in the event of a fire. There have been many reports of CO alarms going off before the smoke alarm in the case of a home fire where the fire was generating tremendous amounts of CO before sufficient levels of smoke were detected.

It is very important to remember that the CO alarm detectors available for consumer purchase and listed under UL 2034 may not be the protection needed by certain individuals with existing health problems or conditions. A low level monitor may be preferable.

Monoxide

What Type of CO Alarm Should be Installed?

The recommendation is to install at minimum, an alarm that meets or exceeds the current listing under UL 2034. Under this listing, CO detectors must alarm within certain times for different concentrations of carbon monoxide. Levels of CO for those time periods allegedly equate to safe concentrations for healthy human beings. This standard requires alarms to signal when indoor air contains CO for periods of time that would equal 10% COHb when breathed by healthy individuals. This is referred to as carboxyhemoglobin (COHb) per cent or the amount of CO found in the human blood stream.

Some advocates and consumer groups feel changes in the standards since 1994 have jeopardized their effectiveness and recommend models specifically designed for particularly sensitive populations. These alarms typically read out and/or detect/alarm at lower ranges of CO exposure. They are priced, generally, in the $80 - $179 range as opposed to the $20 range for the standard models.

Do they have the same sensor capabilities as my test instrument?

The comparatively inexpensive home alarms use sensor technology not equipped to measure and display low level, short term concentrations of CO (usually less than 60 PPM or 100 PPM). The sensors used in Bacharach instruments are highly responsive to low levels of carbon monoxide in the sampling environment.

The typical home alarm will not read out (nor alarm) at lower levels of CO that may be present and displayed by a CO test instrument. Nor will they react as fast as an instrument.

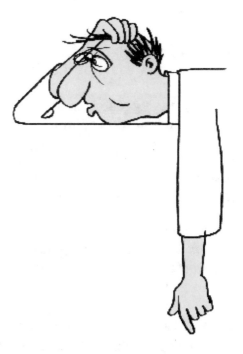

They are not designed to do so.

Remember, a more vulnerable human being such as an infant or someone with congestive heart problems or breathing problems may suffer chronic ill effects from these lower levels.

Carbon Monoxide

When encountering households containing inhabitants other than healthy young adults, a more sensitive CO alarm system should be suggested. They may have a higher installation cost, but are more effective for every body involved in the long term.

It must be noted that home carbon monoxide alarms with digital displays have been reported in the field to have low CO PPM displayed on the CO alarm but 0 PPM or trace measurements inside the building with hand-held, portable instruments. When a hand held CO detecting instrument like the Monoxor II® or CO Snifit® is used, this discrepancy is often noted.

CO alarm sensors can also be affected by or cross sensitive to a number of other factors. Temperature, humidity, calibration requirements, and sensitivity to low level, real time measurement for both types of devices are different. Read the instructions. The typical home alarm, even those with a more sophisticated data logging and peak measurement recorder, do not protect all inhabitants from potentially harmful concentrations of CO.

There have been reports of CO alarms responding to a number of common household compounds. The following is a list of common household chemicals and other substances that may have an effect on the sensor:

- Aerosols–(hair sprays, deodorizers, etc...)
- Cleaning supplies – (Clorox, bleaches, etc...)
- Gas from charging batteries
- Paints
- Stripping chemicals
- Varnish
- Silicon glue or compounds
- Alcohol
- Methane
- Toluene
- Acetone
- Nail polish
- Nail polish remover
- Sulfur compounds
- Sewer gas
- Vapors from baby diapers
- Car exhaust fumes
- Cigarette smoke
- Incense smoke
- Ammonia
- Carpet cleaning solutions
- Sealant
- Freon from air conditioners
- Hydrogen
- Nitroglycerin (usually from heart medication)

It is important to check with the manufacturer of a particular alarm or detector to determine potential cross sensitivity to these or other compounds.

The only gas which test instrument CO sensors are cross sensitive to is hydrogen from, for example, charging an automobile battery in the vicinity.

You may find carbon monoxide levels higher than outside levels and the alarm has not sounded. You may not find measurable CO inside a building where an alarm has sounded. You may feel as if you are chasing ghosts when tracing CO to its source or sources.

To chase this ghost, you have to know what is being chased and how it is being detected. You also need to determine the UL specification listing of a particular alarm that may or may not be sounding to help evaluate the situation.

Monoxide

Changes in the UL Listings for Carbon Monoxide Alarms

The UL 2034 listing requirements for home alarms are based upon CO concentrations measured in PPM and time of exposure.

Carbon monoxide must be present in specific concentrations for specific times before they will alarm.

This time weighted measurement standard has gone through numerous revisions.

As a carbon monoxide source investigator, service technician, or other provider, you may encounter any one of the following three listings.

1. April 30, 1992, UL 2034 listed CO alarms had to measure and alarm when CO is:

15 PPM for 8 hours before alarming, or
100 PPM for no more than 90 minutes before alarming, or
200 PPM for no more than 35 minutes before alarming, or
400 PPM for no more than 15 minutes before alarming.

During this time (prior to 10/95), many communities experienced an increased use of CO alarms by consumers, particularly Chicago. First responders and Authorities of Jurisdiction had little experience with carbon monoxide measurements at these "low" levels. There were suspicions of the accuracy of these alarms and many encounters with CO alarm responses had the responders crying

"False alarm!"

We must examine and remember that CO concentrations outside of buildings in many urban areas will exceed 15 PPM for over 8 hours due to automobile influences. We must also remember the early CO alarm responders most likely were using CO test instruments that auto-zeroed when turned on outside (inside or anywhere). Others used the 'stain tube' type instruments which were not nearly as accurate or responsive.

First responders who had instruments that allowed a manual zeroing of the instrument were doing so inside, close to their vehicles or outside buildings in urban areas where background readings of CO were not zero.

Consequently, many first responders or source investigators were starting with a false zero, thinking there was no CO in the atmosphere they started in, when there was in fact enough to trigger the alarm. Their instruments might not have shown but slight increases of CO upon entering those buildings, so they thought the alarms were inaccurate or defective. Additionally, some of the home alarms sounded off at 9 PPM or lower before 8 hours. This was and still is particularly troublesome for first responders in cities with high pollution.

Another important thing to know about this generation of CO detector is its response criteria for non-alarm status when selected vapors and gases are present in specific concentrations. This Selectivity Test with these substances is intended to represent air contaminants likely to be found in the vicinity of an installed detector.

Methane	500 PPM
Butane	300 PPM
Heptane	500 PPM
Ethyl Acetate	200 PPM
Isopropyl Alcohol	200 PPM
Carbon Dioxide	1000PPM

This low PPM requirement for Carbon Dioxide (CO_2) concentrations may have been another major contributing factor in many 'nuisance' alarms. First responders and other CO source investigators most likely were not carrying any type of CO_2 measuring instruments that could have helped discover the cause of the alarm.

It is not uncommon to find CO_2 levels in residential buildings much higher than 1000 PPM. Given our efforts to save energy, we seal up our homes and businesses. Expired breath alone from occupants in a sealed structure can easily result in a CO_2 build-up over 1000 PPM.

Additionally, unvented combustion equipment, like gas cooking ovens and burners, space heaters, fireplaces and others, dump tremendous amounts of CO_2 (and water vapor) into a space. Under these circumstances, 2,000 to 3,000 PPM or higher of CO_2 in air is not uncommon.

Because of the seemingly high number of false alarms, much mistrust in this first generation of UL 2034 inexpensive home CO alarms resulted. However, despite the number of false or 'nuisance' alarms, many lives were saved from CO poisoning.

Also, many technicians, first responders, fuel suppliers, home inspectors and others began to examine and question their own practices and prejudices about carbon monoxide, as well as codes, licensing, education and other aspects of their profession. It seemed that many of the rules of thumb pertaining to CO had come about with little or no actual field-testing and were being passed around as hearsay.

Many who had been conducting CO and combustion testing prior to the home alarm invasion knew it was a real problem in the field. Their suspicions were that false alarms most often meant inaccurate testing by personnel who did not fully understand CO. Nor were they looking at the building as a system.

Monoxide

After Oct. 1995, UL 2034 listed CO alarms had to measure and alarm when CO is:

15 PPM for no less than 30 days, or
100 PPM for no more than 90 minutes before alarming, or
200 PPM for no more than 35 minutes before alarming, or
400 PPM for no more than 15 minutes before alarming, **and**

Alarms had to have a **reset button** that alarmed if 100 PPM or more was present for at least 6 minutes. In essence, if the alarm sounded and the inhabitants of the building showed no sign of CO poisoning (nausea, passing out, general weakness of the body, etc.), they were instructed to enact the reset. If it went off again, they were to vacate the building and call an Authority. The selectivity Test criteria remained at their same levels.

3. After Oct. 1, 1998, UL 2034 listed CO alarms must measure and alarm when CO is:

30 PPM for 30 days
70 PPM for no more than 189 minutes before alarming (may alarm as early as 60 min.)
150 PPM for no more than 50 minutes before alarming (may alarm as early as 10 min.)
400 PPM for no more than 15 minutes before alarming (may alarm as early as 4 min.) **and** have a <u>manual reset</u> that will reenergize the <u>alarm signal within 6 minutes</u> if the CO concentration remains <u>at 70 PPM or greater</u>.

Another significant change to the 10/1/98 CO alarm listing is the addition in the instructions stating that **individuals with medical problems may consider using warning devices which provide audible and visual signals for carbon monoxide concentrations under 30 PPM.**

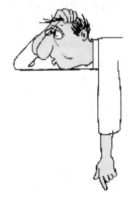

Before you make a recommendation concerning CO alarms, it would be a good idea to know <u>who is in the building.</u> Perhaps a more sensitive alarm with lower CO concentration settings would be more appropriate. **Please install one that meets the needs of all people in the building. Read the instructions and install them to the manufacturers specifications. Please note the health advisory listed on all alarm packages.**

The "Required WARNING" on every UL 2034 Listed CO Alarm states:

"Pregnant women, infants, children, senior citizens, persons with heart or respiratory problems, and smokers may experience symptoms at lower levels of exposure than noted. Individuals with medical problems may consider using warning devices which provide audible and visual signals for carbon monoxide concentrations under 30 PPM."

"WARNING: This product is intended for use in ordinary indoor locations of family living units. It is not designed to measure compliance with Occupational Safety and Health Administration [OSHA] commercial or Industrial standards."

Carbon Monoxide

It must also be noted that Oct. 1, 1998 UL 2034 Residential CO Alarms must meet **the specificity test referencing <u>non-alarm status</u> at specific concentrations of certain vapors and gases.** These concentrations include:

- Methane - 500 PPM
- Butane - 300 PPM
- Heptane - 500 PPM
- Ethyl acetate - 200 PPM
- Isopropyl alcohol - 200PPM
- Carbon dioxide (CO_2) - 5,000 PPM

Please note that the CO_2 concentration was raised dramatically. A 4,000 PPM increase is significant and should result in less CO_2 concentration caused alarms. However, levels of CO_2 in the 2,000 to 4,000 PPM range can be harmful to even the healthiest of individuals and usually indicates lack of air change, combustion gas spillage or both.

We may still encounter buildings where these concentrations exceed these test limitations, particularly CO_2 buildup from lack of ventilation, improper use and placement of unvented space heating or cooking equipment, backdrafting or spillage from vented combustion appliances or systems. Many of the buildings we enter have been over sealed to save energy.

Carbon dioxide (CO_2) is also measured in PPM.

Excessive CO_2 in the air may cause illness symptoms similar to those of CO such as drowsiness, sinus stuffiness or breathing difficulty. These symptoms can be compounded by warmer room temperatures. **A CO_2 and CO measurement** upon entering buildings or rooms provides immediate information about indoor air quality, safety and can be vital to complete CO alarm investigations.

CO_2 LEVELS OF COMFORT IN PARTS PER MILLION

Normal outside levels	350-450 PPM
Acceptable levels	less than 600 PPM
Increased complaints of stiffness and odors	600 to 1000 PPM
ASHRAE and OSHA standards	1000 PPM
Increased complaints of general drowsiness	1000 to 2500 PPM
Adverse health affects	2500 to 5000 PPM
Maximum allowable concentration for 8 hour period	5000 PPM

A technician measuring CO_2 when entering a building may discover whether or not increased ventilation is required or if an unvented appliance is adding unhealthy amounts of carbon dioxide to the living space.

Carbon Monoxide

Why Wait for the Alarm or Injury?

More and more buildings have inexpensive CO detectors that sound off, increasing source investigations. Training and proper use of test equipment are helping to improve the quality of those investigations and can help reduce the number of alarms.

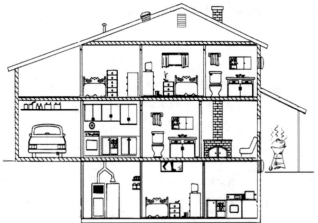

Proper testing performed routinely can increase opportunities for detecting problems before they become emergencies or alarm calls.

Service personnel from the emergency responders, home and building inspectors, to HVAC installers and estimators, and others should use consistent diagnostic procedures when entering a building. The use of a checklist helps ensure consistency. Most importantly, all of our checklists should have a recognizable standard, code or some other obvious or recognizable standard of consistency.

It is vital that we know the CO and code standards for each community we work in because they may be somewhat different for each. If standards or codes do not exist the liability rules of best practices apply. Remember that the consumer may have limited knowledge or understanding of these codes or standards.

We measure CO outside before we enter and, upon entering, determine if there is too much CO and perhaps where it is coming from. We may even fix the problem by using the help of combustion gas analyzers for CO and efficiency. At minimum, we make the proper referral. Additionally, we must have an understanding of carbon monoxide dynamics and how much is too much?

Manufacturers of testing equipment often speak with local Authorities of Jurisdiction, fuel suppliers, professional associations and others in attempts to establish reasonable standards for testing consistency. This program continues that effort.

Carbon Monoxide

The most valuable resources on code adherence for combustion systems are: manufacturers of systems, state, municipal or county or provincial mechanical inspectors, utility companies that are involved in the inspections and fire-up process, workplace regulators, air quality regulators or fire departments.

Always consult with the Authority Having Jurisdiction about:

- **How much carbon monoxide is too much** in homes and what are your responsibilities for testing for it, reporting it, fixing it, and/or notifying the consumer, landlord or other responsible party?

- How much carbon monoxide is allowed to be produced in flue gas by: furnaces, water heaters, ovens, gas or wood fireplaces, clothes dryers, unvented heaters, space heaters, others?

- What are the approved methods for sealing test holes in vented systems?

- Are there any local codes for the work place in reference to CO concentrations or confined space entry?

- What are the licensing and continuing education requirements for my profession within this jurisdiction?

COOPERATION

Cooperation and consistency of test procedures are beneficial to this effort.

Carbon Monoxide

Code Compliance

♦ Does your community employ the use of hazard or warning tags?
♦ Do we understand reporting procedures for these issuances?
♦ How are these inspected or followed up on?
♦ Does your community have a method of inspection for consumers who do their own work?
♦ Does anybody monitor who is installing combustion equipment?

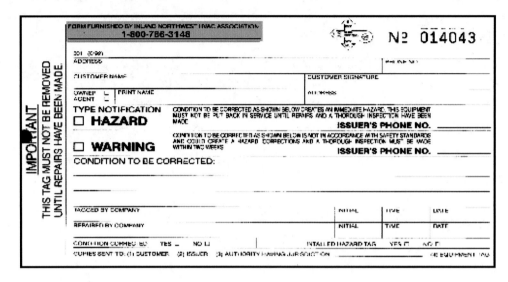

A reporting form or tag, with thorough follow-up, is a good way to help ensure problems are identified and have a reasonable method of repair and re-inspection. Problems associated with carbon monoxide may require reporting forms with greater detail and always require calibrated test instruments sensitive to carbon monoxide.

To measure CO, we have to know how our test instruments work. How often do we need to calibrate them? How do they measure? Do they have any cross sensitivity? Do they auto zero or manually zero? Which is better for my application?

We enter buildings in our business and service for many reasons besides in response to CO alarms. We may be missing opportunities beginning the moment we enter. Continuous measurement during every service visit may reveal conditions in the building and about the building that warrant further testing methods, including breath analysis for CO and data logging for CO, CO_2 or other measurable tests.

We may even find out upon entering a building or a room in the building that the air is not safe or legal for us to be in. *If we don't test and measure we don't know. If it's not test work it must be guesswork.*

Carbon Monoxide

Let's stop and look at this preventative approach to carbon monoxide from a community perspective. Within most communities there is a small army of professionals taking and measuring vital air quality samples and making safety referrals that are based upon local codes and reasonable practice standards.

In any one community, how many people from the HVAC industry are in how many buildings everyday?

How many estimators, installers, service technicians and sales people from how many companies have opportunities for measurement, but don't measure?

For many, measurements taken may also result in additional work. For everyone it is an opportunity to randomly save a life or to effect the poor health condition of some consumer in your community. We may even learn a little bit more about carbon monoxide.

Everyday, how many homes are being entered by home, building or mechanical inspectors? How many public and private sector energy and weatherization program personnel are entering buildings everyday?

How many buildings do fire department, emergency responders, gas utility or other representatives get into daily, weekly, monthly? How many boiler rooms are entered everyday? How many crawl spaces or attics with combustion systems in them are being entered everyday?

We represent a very significant number of consumers. It is our duty to alert or warn and inform them to hazards and to encourage safety and use of recommended procedures for service and repair of combustion systems. It is dutiful to recommend CO detector alarms and regular, consistent service.

Consistent service means measurement and documentation of those measurements. Old rules of thumb have given way to precision. Professionalism succeeds guesswork.

DOCUMENTATION IS IMPORTANT!

Carbon Monoxide

Documentation

The following is an example of a typical worksheet used by HVAC technicians in the field. This form does not provide information necessary for determining CO or CO_2 levels. (For appropriate forms see **Appendix A**).

AAAA Heating and Air Conditioning
123 Any Street
Any City, USA 00550
TELE: 1(555) 555-1000   FAX: 1(555) 555-1004

Date: 09/02/2002	Make & Model: Triple Star 75000
Complaint: Fall Furnace Check and Clean	
Technician: Al W. Aysdoinit	
Customer Name: Mr. John Smith	
Address: 1642 Carbon Street	
City: Any City **State:** USA **Zip:** 00551	
Telephone:	

Qty	Item	Description		Total
	✓	**Thermostat**	Hall adjusted & cleaned	
	✓	**Motor**	dirty	
	✓	**Gas Control Valve**	W/R newer	
	✓	**Igniter / Pilot**	good light off	
	✓	**Thermal-couple**	replaced	
	✓	**Efficiency**	rated to furnace	
	✓	**Heat Exchanger**	no cracks	
1	✓	**Filter**	needed new one 20 X 20 X 1	
	✓	**Cleaning**	yes	
	✓	**Duct Work**	good	
	✓	**Gas Leaks**	OK	
	✓	**SAFETY &EFFICIENCY**		
		TOTAL		$69.99

Carbon Monoxide

General Heating System Inspection Form	System #1	System #2	Comments
Carbon Monoxide (CO) levels in outside air	Yes ___ No ___ If yes, ____ PPM		
CO level upon entry into building.	Yes ___ No ___ If yes, ____ PPM		
If present, has the CO detector ever alarmed?	Yes ___ No ___		
Is the installation of a CO alarm recommended?	Yes ___ No ___		
Combustion system type?	FA BO SH H_2O OV FP OT	FA BO SH H_2O OV FP OT	
Manufacturer/ model #			
Are there excess CO levels in flue gases during steady state operation?	Yes ___ No ___ If yes, ____ PPM	Yes ___ No ___ If yes, ____ PPM	
Type of fuel?	NG LP FO OT	NG LP FO OT	
Gas/oil leaks identified?	Yes ___ No ___	Yes ___ No ___	
If yes, describe location.			

There are many examples of reporting forms. Many have similarities in the initial steps of source investigation.

As we explore more of the carbon monoxide phenomenon, our role in this effort becomes clearer.

Carbon Monoxide Safety Checklist

Occupant Identification and evaluation: Date: _____

Name: _____ Phone: (____) _____

Address: _____ City/ST/Zip _____

Are any occupants symptomatic? Yes ___ No ___

Indicate presence of symptoms below and the number of people effected.

___ Headache # ___   ___ Fatigue # ___   ___ Nausea # ___   ___ Confusion # ___

___ Shortness of breath # ___   ___ Other # ___   describe _____

CO alarm? Yes ___ No ___   Sounding? Yes ___ No ___

Digital indication? Yes ___ No ___ PPM ____

Carbon Monoxide

Responding to a Carbon Monoxide Alarm

Dispatch

As soon as emergency rescue personnel or an HVAC company gets a call, trained professionals need to be ready to respond. When the call concerns a home alarm of some type, dispatchers and HVAC technicians around North America have had to determine if it is a:

1. Smoke alarm or a CO alarm.
2. Where the people are calling from (a house on fire is no place for phone calls)
3. Has anyone passed out, vomited or showing any other CO poisoning signs?
4. Does the CO alarm have a reset button? Has it been reset? Are you still calling from the house?

All information should be gathered without jeopardizing anyone's safety. Consistency of documentation through use of the Carbon Monoxide Poisoning First Response Checklist found in this manual can help ensure that consistency.

Instruction must take place immediately as verbal questioning warrants. If they haven't left the building and the alarm has sounded again after reset, they should do so immediately if physically possible.

If one or more of the people in the building have headaches, vomiting or showing any other physical indications of CO poisoning, they should all get to fresh air immediately.

If someone has fainted or is unconscious and/or cannot be moved outside the affected area, windows and doors should be opened, especially in the room where the unconscious victim is. Get everyone outside.

Instruct them *not* to go around opening windows, doors and turning off appliances if everyone can get outside immediately.

Someone in a potentially contaminated environment should not be told to go around opening doors and windows risking further exposure.

Ventilating a structure may make it impossible to determine if CO was indeed present as well as increase the difficulty in identifying the source.

Callers should be advised to go to a neighbor's house if possible. The calling adult should wait for the first response team in front of the building. Complete cooperation with the first responders is vital.

Many times the service department of a regulated gas utility company is notified in direct phone link to all 911 CO alarm calls. The cooperation between the gas utility companies and the fire departments in most areas has been professionally responsive. This same cooperative effort appears to be absent in many rural areas where non-regulated fuel supply exists.

Carbon Monoxide

First Response

There are many sources of CO in a building. When responding to CO, the protection of the inhabitants and all response personnel is primary. The emergency response team must share in and understand the importance of each investigative step. Each team member should have their own test instrument.

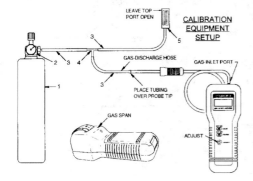

Before testing the buildings' air for CO you must have an instrument calibrated to manufacturers specifications.

If you are using the Monoxor II®, you may manually zero the display. This can only to be done in environments where no CO is present. This adjustment needs to be done every few months. There is little drift in the instruments display if we understand its boundaries of operation and are careful and observant during its use. If not turned on in a CO free area, auto zero instruments will register above this false zero start.

1. Cylinder, Calibration Gas (For CO calibration, use Bacharach P/N 24-0492. Recommended flow is 2-3 liters/minute at approx. 10 PSI)
2. Regulator, P/N 51-2974
3. Tubing, P/N 03-6351
4. Tee, P/N 03-5532
5. Flow meter, P/N 06-6163

Once outside air references are established, record them. If the calling consumer is not waiting in front of the building, the decision to enter becomes immediate: There may be someone inside.

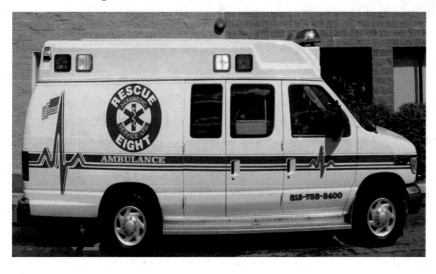

If the caller is waiting outside the building, find out about all other inhabitants' locations and general health while another team member enters the building, CO measuring device in hand.

Before entering any building on a CO or CO_2 alarm call, some emergency responders are required (if it is local department policy) to wear a self-contained breathing apparatus (SCBA). In some jurisdictions it is not uncommon to find SCBA's worn if the inside atmosphere contains over 35 PPM or even 50 PPM.

Once inside, measure air and look for people or animals. Confusion is one of the symptoms of CO poisoning and even though the caller said everyone was out of the building, your duty is to check again. Record inside measurements.

Outside a team member should be with the caller. First address those most sick, administer oxygen and call for back-up support if needed.

If none of the building inhabitants demonstrate debilitating CO poisoning symptoms, obtain a breath sample from whoever has been in the building the longest time and perhaps the healthiest (non smoking) person in the group. Attach the Breath Analyzer Module to the Monoxor II®.

Carbon Monoxide

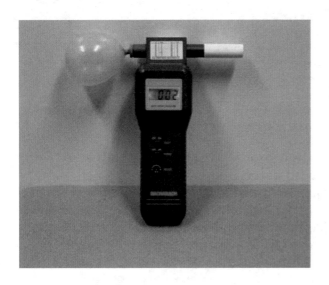

CO Breath Analysis Module							
CO PPM	COHb %	CO PPM	COHb %	CO PPM	COHb %	CO PPM	COHb %
0	1	50	9	120	21	220	37
05	1	55	10	130	23	230	39
10	2	60	11	140	25	240	41
15	3	65	12	150	26	250	
20	4	70	13	160	28		
25	5	75	13	170	29		
30	5	80	15	180	31		
35	6	90	17	190	33		
40	7	100	18	200	34		
45	8	110	20	210	36		

Have the subject inhale completely and hold breath for 15-20 seconds. Subject should exhale about one-half of breath into the atmosphere and then, following instrument instructions, breathe the remaining breath into the mouthpiece of the BAM. The balloon should be inflated in this manner until it reaches approximately 5" in diameter.

The expired breath will move through the test instrument and CO concentration in PPM will be displayed on the digital display. This reading is not COHb %. We have to convert that PPM measurement. The accompanying chart will demonstrate the approximate equivalencies of PPM and COHb%. If a poisoning of any degree is verified, procedures for poisoning response should be followed.

Smokers may record a variance of COHb. This range usually is within 2% to 6% COHb. However, the amount of measured COHb in a smoker depends upon how soon after smoking the sample is taken and how heavy a smoker the victim is. Higher measurements may occur if a test is taken within the first ten minutes after a cigarette as opposed to one hour or more. The amount of cigarettes per day and smoke actually inhaled also has an effect on this measurement. (Often times in Bacharach seminar, a smoker volunteers for this breath test and the measurement is above 6% and not unusual to be over 10%)

It is suggested that samples be taken from several people in the building. Record all measurements.

People living in urban areas where auto exhaust is high and background CO measurements regularly record above 9 PPM may also have breath samples around 2% even if they are non-smokers. The human body produces some CO naturally. About ½ to 1% is a normal reading for a non-smoker who lives in fresh air. If the average healthy person inhales 70 PPM for around 3 hours, a measurement should find around 10% COHb.

Carbon Monoxide

Once you have quantified the CO health of all people, your next steps are determined by the supporting activities of fuel suppliers and other responders. In some rural areas, the first responders must also perform CO source investigations and test individual appliances within the building. If you have source investigation support by qualified technicians, your paperwork and possible victim placements are all that is required.

There are many scenarios for each situation. If windows or doors were left open, the house may have ventilated before you arrived. Your interior measurements may indicate little or no CO. It is always recommended that appliances be tested before you allow the occupants back into the building.

Breath analysis may confirm the presence of CO, but measurements inside may indicate no CO. Complete testing of the building must be performed. This will take time. The housing of the inhabitants is then of concern. The decision then is how, when and by whom is the next set of tests to be performed and who pays for it.

Leaving a Bacharach Comfort Chek® with CO data-log capabilities in the building for several days will provide additional information about times of day when CO is highest in the building under normal operating conditions. Noting the higher concentrations, when they occurred and finding out what systems were running, what weather conditions existed, how many people were home and other information can enlighten the source investigation team.

If we don't measure, we don't know.
CO standards and guidelines
What standards do I use if local codes do not address this issue?
REMEMBER:
Two circumstances must exist to make carbon monoxide a hazard:

- It is produced in concentrations that can affect health or hurt someone.
- An open path exists for CO to reach people.

Carbon monoxide has a natural tendency to rise in temperatures normally found inside buildings. At freezing temperatures, carbon monoxide is heavier than air. Please keep in mind that cold days in areas of high automotive activity, outdoor ambient CO levels may be more noticeable than on days above freezing.

Other combustion by-products can contribute to poor health symptoms even when CO is not produced. These by-products include NO_x gases and excessive CO_2. Acidic moisture from combustion gases can also be harmful.

Though carbon monoxide is odorless, a distinct pungent odor can occur during incomplete combustion. Aldehydes are a by-product of incomplete combustion and come with this pungent odor. Often times this odor is mistaken by the consumer as a gas leak. Auto exhaust may also have an odor of raw fuel.

Carbon Monoxide

The deceiving part about carbon monoxide is that it does not have an odor.

Carbon monoxide is given off by the incomplete burning of solid, liquid or gaseous fuel. This occurs when there is not enough oxygen mixed with the fuel and/or proper combustion conditions are not met. Variations in combustion conditions can result in varying levels of CO.

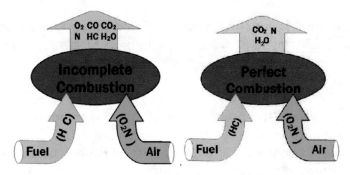

What is the fuel being used? How many BTU's per cubic foot or how many BTU's per gallon? We have to know the fuel.

VERIFYING PROPER COMBUSTION CONDITIONS INCLUDE:
1. The measurement of fuel flow: Measure in water column inches or PSI (Pounds per square inch). Always check manufacturer specifications. Flow of fuel is specific, measure it.
2. Verify that combustion air is adequate, to code and does not interfere with combustion systems performance. Air and fuel mixtures are measurable. Combustion analyzers offer O_2, CO and flue-stack temperature measurements. CO_2 efficiency and CO air free are calculations.
3. Verify that systems with vents will draft without interruption and are installed in accordance to vent manufacturers sizing, installation and performance requirements.

Fuel burns completely, without impingement or interference – minimal CO production.

If there is a CO problem, is it identifiable and is it correctable? Do we need to vacate the premises? Do I have a duty to report this level to an Authority?

Our procedural steps should include rating the levels found and referencing them to reasonable and accepted levels. Many standards for CO exposures come from nationally recognized professional associations or governmental administrative organizations. The medical community has helped establish these levels. Standards change as new information becomes available to the industry. Check **Appendix B** for a list of contacts on current standard information.

Many local Authorities of Jurisdiction have **living space** limits for carbon monoxide. It is recommended you ask the local fire department what their limits are for evacuation. In some locations the regulated gas utility may be the Authority. Otherwise the city, county, or state inspection department may be the authority.

Carbon Monoxide

How Much Carbon Monoxide is too Much?

Always check with The **Authority Having Jurisdiction**. Know who is in the building being tested.

ASHRAE 62-89 (American Society of Heating, Refrigeration and Air Conditioning Engineers)
009 PPM: The maximum allowable concentration for continuous (24 hr) exposure. ASHRAE states the ventilation air shall meet the out door air standard referenced to EPA and 9 PPM.

EPA (Environmental Protection Agency)
009 PPM: This level or lower as an ambient air quality goal averaged over eight (8) hours. This out door air standard is exceeded in many urban areas due to auto exhaust.

Common Action Level
009 PPM or more above what you measured outside is the most common action level in the U.S. by local Authorities of Jurisdiction for further testing. Some jurisdictions require fuel shut-off until problem diagnosed and corrected.

BPI (Building Performance Institute)
10 to 35 PPM is a marginal level in reference to potential or foreseeable problems in some situations. Occupants should be advised of a potential health hazard to infants and small children, elderly people and persons suffering from respiratory or heart problems. If building has attached auto garage, document CO levels in garage. Accept this level as normal where unvented appliances are in use. These levels are unacceptable when originated from vented appliances.

UL 2034 (Underwriters Laboratories, Carbon Monoxide alarm detector designation)
30 PPM is the concentration required for UL 2034 listed alarms to sound when this concentration is present for 30 days minimum. This allows the sensor to clear itself. People of vulnerable health may require alarms with lower PPM concentration trigger levels.

EPA
035 PPM: This level or lower as an ambient air quality goal averaged over one (1) hour.

Common Action Level
035 PPM is also a common action level for fire department or other emergency responders to utilize self contained breathing apparatus when occupation of that environment is to be sustained by that responder.
035 PPM or less averaged over an 8 hour day within that workday is a common goal of specific States Occupational Health and Safety Administration or similar state entity. This is also a common goal of many employers despite higher regulated concentration standards and may require the measurement of several simultaneous reference locations.

OSHA (Occupational Health and Safety Administration)
050 PPM: Maximum allowable concentration for a workers continuous exposure in any eight (8) hour period. This 8-hour average requires continuous measurement and accurate reporting in the workplace.

UL 2034 - 98
70 PPM concentration required for UL 2034 listed CO alarms to sound when concentration is present for no more than 240 minutes (4 hours) or as early as 60 minutes (one hour).

Carbon Monoxide

BPI
36 to 99 PPM is excessive. Medical alert. Conditions must be mitigated. Ventilation required. Always test garage space. Individually test combustion appliances. All repairs are to be conducted by a qualified technician with proper test equipment.

BPI
100 to 200 PPM is dangerous (**and is a common building evacuation standard.**) Medical alert conditions exist. It is suggested that occupant health inquiries be conducted. It is advisable that someone else transports them to seek medical help; 15 minute maximum exposure upon discovery. Report incident to Authority of Jurisdiction.

UL 2034 - 98
150-PPM concentration required for UL 2034 listed CO alarms to sound when concentration is present for no more than 50 minutes or as early as ten minutes.

200+ PPM is extremely dangerous. Universally accepted as an evacuation action level. The health of occupants should be monitored and emergency conditions may exist. Building should be ventilated and searched for additional occupants. Combustion systems should be thoroughly tested for CO production and dispersion. Report incident to an Authority of Jurisdiction.

UL 2034 - 98
400-PPM concentration required for UL 2034 listed CO alarms to sound if concentration is present for no more than 15 minutes or as early as 4 minutes.

Any increase in PPM from outside to inside warrants further source investigation and is documented, reported and even fixed is common in jurisdictions where a fuel supplier also is considered an Authority of Jurisdiction. This standard is also common to some federally and state funded weatherization programs as well as protocol to some private companies engaged in carbon monoxide testing.

CO Air Free Standard

Carbon monoxide measurements in flue gases also must meet specific concentration standards. Though CO concentration standards have existed for appliances and other combustion systems for many years, lack of testing and misunderstanding of CO measurement has occurred.

As stated earlier, carbon monoxide can be measured in flue gases as CO PPM **or** CO PPM Air-free. Appliance manufacturers must produce units that comply with measurements of CO Air-free that are less than the listed maximums.

The following CO PPM Air-free measurements are offered along with common CO PPM standards found in jurisdictions where single sensor CO instruments are used.

ANSI Z21 (American National Standards Institute)
200 PPM CO Air-free is the maximum concentration from an unvented space heater, vented gas water heater.

ANSI Z21 and EPA
400 PPM CO Air-free is the maximum allowed in furnace flue gas.

ANSI Z21
800 PPM CO Air-free is the maximum allowed for gas oven emissions.

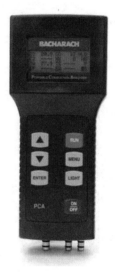

To many technicians who have been performing combustion gas tests regularly, these concentrations are extremely high in reference to CO levels that can be expected from combustion systems finely tuned and maintained.

CO Air-free measurements are not always performed due to lack of understanding about the requirement and lack of technicians' understanding about the importance of O_2 measurement in establishing the efficiency and safety parameters of combustion systems.

Carbon Monoxide

This manual has discussed air measurement standards for CO in ambient air and flue gas standards for CO Air free.

The most common criticisms about flue gas testing often comes from technicians or companies that do not test. Some of the most common arguments against testing include:

- *"If I test, I become liable."* and
- *"The instruments cost too much to buy and maintain."*

Fortunately, many technicians and companies regularly test flue gases on furnaces, water heaters, ovens, boilers and other systems. They are aware that liability begins as soon as they walk in the door, and that testing generates more business and actually reduces liability. Testing generates additional profits, more than offsetting the cost and maintenance of equipment.

Carbon monoxide problems can be identified and minimized.

It is common to find technicians, gas utility company personnel and others who do not understand the CO Air Free measurement and who often mix Air Free and non-Air Free measurement numbers.

What if I use an instrument with only a CO sensor and can't measure O_2 & calculate for CO Air Free?

Many technicians and inspectors use single gas CO test instruments (like the Bacharach Monoxor II® or Monoxor III®). The action levels they use for how much CO is too much in flue gases is reported in CO PPM not CO PPM Air Free because the excess air in that flue gas sample has not been calculated out of the reading.

The CO in PPM will be a significantly lower number than the CO Air Free measurement ceiling offered by ANSI, EPA or AGA.

Knowing that an atmospheric natural gas system's flue gases generally contain 7 to 9% O_2, and using our CO Air Free formula, we would calculate the furnace ceiling amount of 400 PPM CO Air Free to be equivalent to around 225 PPM CO or higher when using a single gas instrument. However, without measuring the contenet of the flue gases, the precise CO Air Free calculation cannot be verified.

(See pages 3 and 4 to reference the CO Air Free formula)

Carbon Monoxide and Combustion Testing Procedures

It is important to know and work within the local Authority of Jurisdictions standards for CO concentration limits in flue gas. Concentrations are referenced to a steady-state or stabilized condition of the system's operation. Sampling points for combustion gas are located before the draft hood of an appliance or other entries of dilution air. Test procedures and conditions are examined in the following sections of this manual.

<100 PPM CO air free
Gas furnaces, space and water heaters are usually considered safe and left in operation. It is reported to be as low as 25 PPM in some weatherization programs. The more often testing is performed, the more the technician understands what is achievable and reasonable.

<150 PPM CO air free
Common ceiling concentration for unvented gas oven emission. Repair recommended if over this amount.

100 - 400 PPM air free
Gas furnaces, space, water heaters and boilers require further testing and correction. Not necessarily immediately lethal concentrations of CO but conditions generally found to be correctable within parameters of normal service work. Systems are generally left in operation with set time limits for correction to be enacted.

400+ PPM air free
This is above ANSI concentrations for flue gas.
Frequently, gas systems are shut off and/or corrected when levels exceed this concentration.

Carbon Monoxide Sample Locations

(The following information section applies to, and is repeated in the Combustion section)

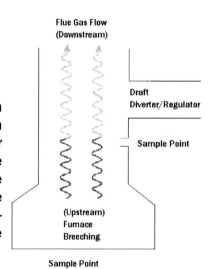

The measurement for gases and temperature should be taken at the same point. This is done by selecting a sample location 'upstream' from the draft diverter/hood, barometric control or any other opening, which allows room air to enter and dilute flue gases in the stack. In larger installations it may also be necessary to extract a number of samples from inside the flue to determine the area of greatest flue gas concentration. Another common practice is to take the flue gas sample from the 'Hot Spot' or the area with the highest temperature.

Make sure that the sample point is before any draft diverter/hood or barometric damper so that the flue gasses are not diluted.

The sample point should also be as close to the breech area as possible. This may also provide a more accurate CO reading should air be entering the flue gas stream through joints in sheet metal vent connectors.

Carbon Monoxide

Oil Burners - Locate the sampling hole at least six inches upstream from the breech side of the barometric control and as close to the boiler breeching as possible. In addition, the sample hole should be located twice the diameter of the pipe away from any elbows.

Gas Burners - Locate the sampling hole on power burner fired boilers/forced air units at least six inches upstream from the breech side of any double acting barometric control and as close to the boiler breeching as possible. Again, try to stay away from elbows. When testing atmospheric equipment with a draft diverter/hood, the flue gas sample should be taken inside the port(s) where flue gases exhaust the heat exchanger.

Carbon Monoxide

Is it measurable? Did it get measured?

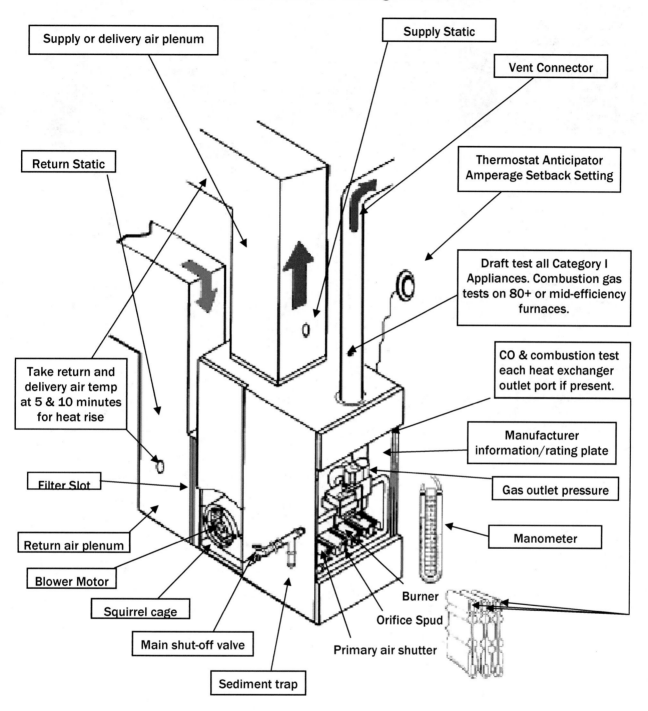

Manufacturers may require heat rise measurements in some other location. Watch for the air conditioning coil when placing test hole and thermometer.

Are Your Measurements Within Manufacturer's Specifications???

Carbon Monoxide

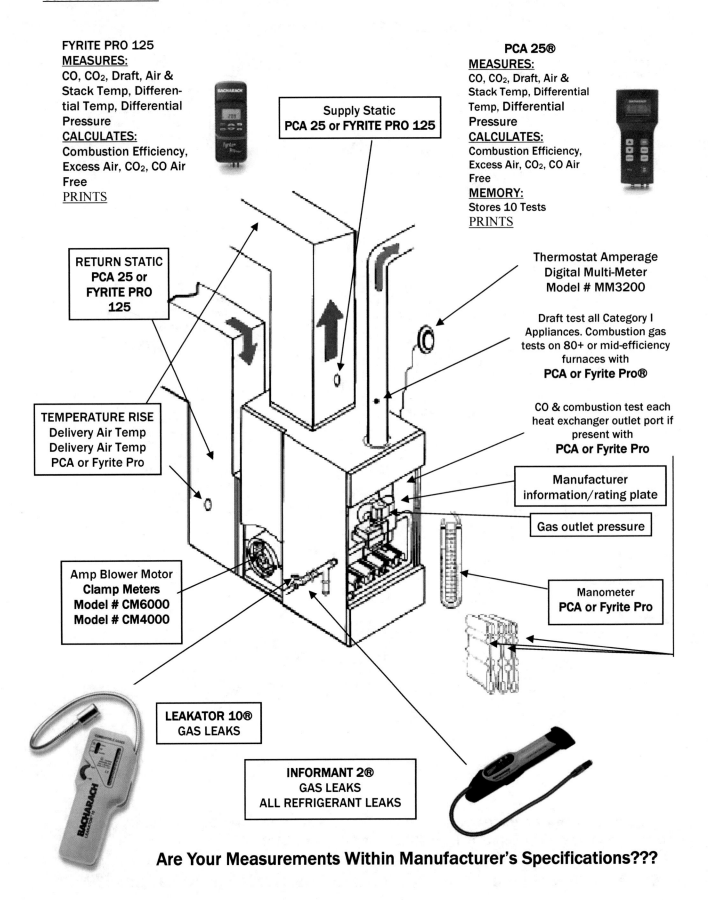

Are Your Measurements Within Manufacturer's Specifications???

Review Questions
(answers on page 54)

1. Carbon monoxide is a result of;
   A. furnaces over heating.
   B. incomplete combustion.
   C. impurities in fossil fuels.
   D. moisture in flue gases.

2. Carbon monoxide;
   A. can be associated with an odor and smoke.
   B. can be identified with a yellow flame.
   C. is odorless tasteless and colorless.
   D. is seldom a problem with fuel oil combustion systems.

3. When a pan of cold water is placed on a range top burner CO maybe produced due to the following phenomenon:
   A. Unburned fuel is pushed out of the flame.
   B. A dramatic cooling of the flame occurs.
   C. The pan causes a change in the combustion environment.
   D. All of the above.

4. CO Air Free is the unit of measurement designed to compensate for;
   A. high fuel pressure and combustion air.
   B. low draft and excess fuel.
   C. excess air in the flue gas samples.
   D. excess levels of combustion air in the plenum.

5. Carbon monoxide inhaled into the lungs bonds with;
   A. cholesterol.
   B. triglycerides.
   C. hemoglobin .
   D. allergens.

6. Children and small pets may exhibit symptoms of CO exposure;
   A. sooner that adult humans.
   B. after adults humans.
   C. at the same time as adult humans.
   D. Not at all.

7. A Carbon monoxide test should be performed;
   A. when a CO poisoning is suspected.
   B. when gas burning appliances are equipped with a draft hood.
   C. in commercial buildings.
   D. anytime there is any possible source of CO production.

Carbon Monoxide

8. The use of work sheets and check lists during routine inspections and service calls will;
   A. ensure consistency and a level of professionalism.
   B. not affect the work done by the HVAC company.
   C. change the refrigeration charge in the A/C system.
   D. change the temperature rise in the heating system.

9. Standards for CO exposure come from;
   A. nationally recognized professional associations, government administrative organizations, medical community and local authorities having Jurisdiction.
   B. British Thermal Unit criteria.
   C. Appendix B.
   D. None of the above.

10. Proper combustion conditions for hydro-carbon fuel systems include:
    A. Acceptable heat exchanger leakage in furnaces with no gas or oil leaks.
    B. Perfect combustion in BTU per cubic foot with PSI verification.
    C. Proper fuel pressure with fuel burning completely without impingement or interference.
    D. Verifying odors in ducted systems with reference to soil stack and CO Air Free measurements in ambient air.

Answers
1 2 3 4 5 6 7 8 9 10
C A D A C D C C A C

COMBUSTION

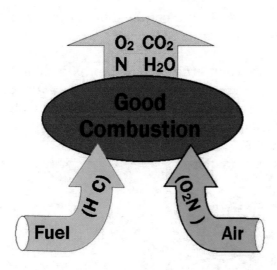

Combustion refers to the rapid oxidation of fuel accompanied by the production of heat, or heat and light. Complete combustion of a fuel is only possible in the presence of an adequate supply of oxygen.

Approximately 1,600 to 2,000 cubic feet of air is required to burn one gallon of #2 fuel oil at 80% efficiency (at sea level). About 15 cubic feet of air is required to burn one cubic foot of natural gas at 75% efficiency (at sea level).

Oxygen (O_2) is one of the most common elements on earth making up 20.9% of our air. Rapid fuel oxidation results in large amounts of heat. Solid or liquid fuels must be changed to a gas before they will burn. Usually heat is required to change liquids or solids into gases. Fuel gases will burn in their normal state if enough air is present.

Most of the 79% of air (that is not oxygen) is nitrogen, with traces of other elements. Nitrogen is considered to be a temperature reducing dilutant that must be present to obtain the oxygen required for combustion.

Nitrogen reduces combustion efficiency by absorbing heat from the combustion of fuels and diluting the flue gases. This reduces the heat available for transfer through the heat exchange surfaces. It also increases the volume of combustion by-products, which then travel through the heat exchanger and up the stack faster to allow the introduction of additional fuel air mixture.

This nitrogen also can combine with oxygen (particularly at high flame temperatures) to produce oxides of nitrogen (NO_x), which are toxic pollutants.

Combustion

Air for combustion is divided into four types depending upon its role and the design of the particular burner. Air will be referenced in this manual as primary, secondary, excess and dilution air.

Primary air provides a percentage of the combustion air, but more importantly, controls the amount of fuel that can be burned.

Secondary air improves combustion efficiency by promoting the fuel to burn completely. Power burners generally do not require secondary air. However, air leaking in through access/clean out doors, burner mounting flanges, boiler sections, etc., dilutes the flame and flue gas temperatures, reducing operating efficiencies as well as our ability to accurately monitor combustion conditions.

Excess air is supplied to the combustion process to ensure each fuel molecule is completely surrounded by sufficient combustion air. As a burner tune-up improves the rate at which mixing occurs, the amount of excess air required can be reduced.

Dilution air does not participate directly in the combustion process and is primarily required to attempt to control stack draft and reduce the likelihood that moisture in the flue gases will condense in the vent system which directly influences combustion air intake, safety and efficiency.

Combustion

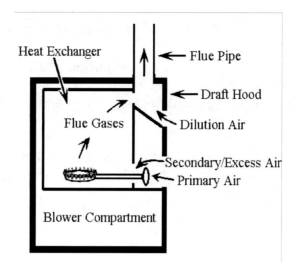

Commonly used fuels such as natural gas and propane consist of carbon and hydrogen, as well as other impurities. When a fuel has a large ratio of hydrogen, more excess air must be provided. Water vapor is a by-product of burning hydrogen. To maintain a vapor state, water vapor absorbs heat from the flue gases, which would otherwise be available for more heat transfer.

Natural gas contains more hydrogen and less carbon per BTU than fuel oils, and as such produces more water vapor. Natural gas is slightly less efficient than fuel oil.

Too much, or too little fuel with the available combustion air may potentially result in un-burned fuel and carbon monoxide generation. A very specific amount of O_2 is needed for perfect combustion and additional (excess) air is required for good combustion. Excess air can contribute to CO generation, lower efficiencies and unsafe conditions. High CO generation can lead to reduced equipment service life.

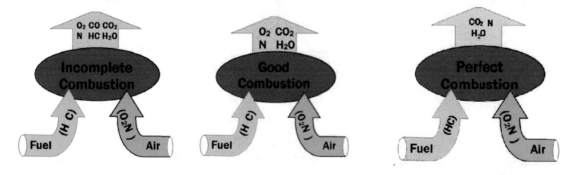

When hydro-carbon fuels are burned, the carbon combines with oxygen from the air and produces CO, CO_2 or both. In perfect combustion (complete burn) CO is not produced.

A number of experiments have shown that when one pound of carbon is burned CO_2 and approximately 14,500 BTU's are produced. When a pound of carbon is incompletely burned producing CO, only 10,200 BTU's are produced. An incomplete burn can result in carbon being deposited on heat exchanger walls or vent systems surfaces, further reducing efficiency and/or increasing safety risks.

Combustion

Principles of Combustible Gas for Technicians

Gases used to supply heat energy are called fuel gases. The major components of fuel gases are hydrogen (H) and carbon (C). The number of carbon and hydrogen atoms linked together determine the weight, heat value and chemical name of the fuel.

```
      H
     HCH
      H
  METHANE (CH4)
```

Methane (CH_4) is the largest component of natural gas and has one carbon atom linked to four hydrogen atoms.

```
     HHH
    HCCCH
     HHH
  PROPANE (C3H8)
```

Propane (C_3H_8) has three carbon atoms linked to eight hydrogen atoms.

```
     HHHH
    HCCCCH
     HHHH
  BUTANE (C4H10)
```

Butane (C_4H_{10}) has four carbon atoms linked to ten hydrogen atoms.

Hydrocarbon fuels contain molecules consisting of hydrogen and carbon. A gas is heavier when more carbon atoms are present in a gas compound. A fuel gas compound containing larger numbers of carbon and hydrogen atoms will release more heat when burned.

Specific gravity refers to the ratio of the density of a solid or liquid to the density of water at 4° Celsius. The specific gravity of a gas is the ratio of one cubic foot of that gas compared to one cubic foot of dry air. The dry air and the gas to be measured must be at the same temperature and pressure.

Gases expand when heated and contract when cooled. The gas industry defines standard atmospheric conditions of pressure and temperature as 29.92 inches of mercury (Hg) and 59°F.

Combustion

In the gas industry, dry air is the reference gas and always has a specific gravity of 1.0. This is a relative number or ratio used to express specific gravity of other gases.

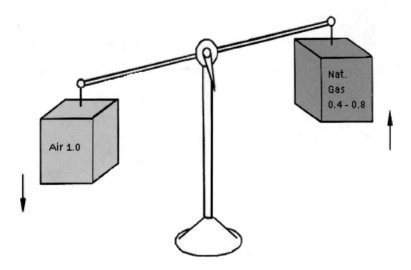

The specific gravity of natural gas ranges from 0.4 to 0.8. This means that a cubic foot of natural gas will weigh only 4/10 to 8/10 that of a cubic foot of air. Natural gas will readily mix with air.

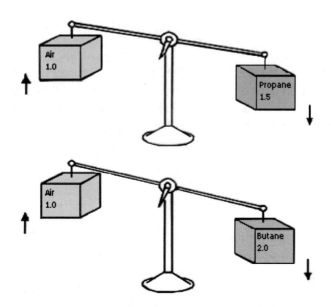

The specific gravity of propane is 1.5. Propane is heavier than air. The specific gravity of butane is 2.0. Natural gas, being lighter than air will usually mix readily with air. Propane and butane, being heavier than air will not readily mix with air.

Combustion

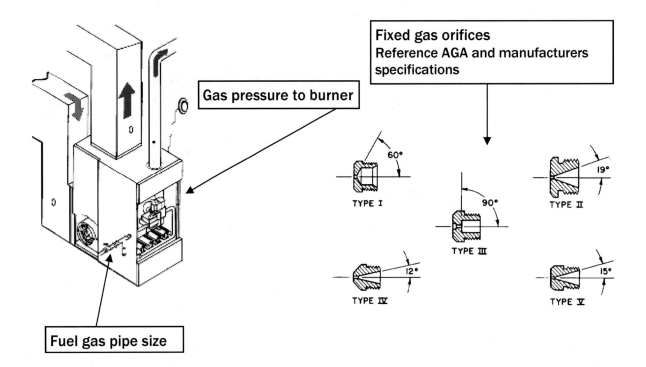

Verify that your Measurements to Manufacturers Specifications!

Specific gravity is taken into account in the selection of orifice sizes for burners, regulator pressure and fuel gas pipe sizes. These sizes and pressures are usually regulated by code. Gas flow rate through an orifice depends on the orifice size and pressure upstream of the orifice. Gases with a lower specific gravity will flow faster through a given orifice size than will a gas with a higher specific gravity. Similarly, a given pressure at a pipe inlet will move a lighter gas with less resistance than heavier gas through that same size pipe.

Fuel gases are commonly delivered to combustion systems in pressures referenced in water column inches. A water column inch (1 WC") is the pressure it takes to push a column of water in a tube up one inch. Optimal delivery for natural gas to a burner is 3 ½ WC". Optimal delivery for propane to a burner is 11 WC".

Natural gas extracted from below ground wells may contain small amounts of condensables. These condensables may be removed as liquids through changes of temperature and pressure. It is important to remove the condensables that contain sulfur as they contribute to the corrosion and breakdown of fuel gas piping and reduce the service life of combustion equipment. Not all condensables can be completely removed from natural gas, trace amounts remain.

Combustion

Liquefied petroleum gases (LPG or LP gas) are propane, butane, or a mixture of the two. These gases are obtained from natural gas or as a by-product from the refining of oil. Trace amounts of condensables also remain in these gases.

Trace amounts of hydrogen sulfide are found in natural gas. Hydrogen sulfide is a condensable that has a corrosive effect on certain metallic surfaces. It is vital that proper fuel gas piping be used to avoid deterioration of these surfaces.

Natural gas, propane and butane are nearly odorless after condensables are removed. Odorants are added to the gas before distribution to aid in leak detection. Some odorants used are colorless liquids containing sulfur compounds that have a garlic like odor, and have become recognized as having a gas-like smell.

The amount of odorant needed to produce a strong, recognizable odor in a gas is quite small. Although sulfur is corrosive, good combustion should burn the minute amount of sulfur in the flame, producing little or no odor or harmful products.

Combustion

Heating Value

Heat energy produced when burning a fuel gas is expressed in British thermal units (BTU). One BTU of heat will raise the temperature of one pound of fresh water one degree Fahrenheit.

1 LB WATER AT 70° F + 1 BTU HEAT = 1 LB WATER AT 71°F

The heating value of a gas is the amount of heat released when one cubic foot of the gas is completely burned and is expressed in BTU per cubic foot of gas at standard pressure and temperature.

The more carbon and hydrogen atoms present in each molecule of fuel gas, the higher the heating value.

Natural gas is commonly referenced as having 900 to 1,150 BTU per cubic foot.
Propane is commonly referenced as having about 2,500 BTU per cubic foot.
Butane is commonly referenced to have about 3,200 BTU per cubic foot.

H HCH H METHANE (CH_4)	HHH HCCCH HHH PROPANE (C_3H_8)	HHHH HCCCCH HHHH BUTANE (C_4H_{10})
900 - 1,150 BTU per cubic foot	2,500 BTU per cubic foot	3,200 BTU per cubic foot

Another unit of heat energy used in the gas industry is the **therm**. A therm is 100,000 BTU of heat energy.

Therefore, if 1 cubic foot of natural gas contains 1,000 Btu, then 100 cubic feet of natural gas is required to produce 1 therm. This is mathematically represented as: 100,000 Btu ÷ 1,000 Btu = 100 cubic feet.

Forty cubic feet of propane contains a therm of heat energy and is mathematically represented 100,000 Btu ÷ 2,500 Btu = 40 cubic feet.

Controlled Gas Fuel and Combustion

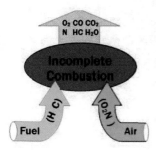

Fuel gases will not burn without oxygen (O_2). Air is composed of about 20.9% O_2, and about 79% nitrogen (N).

O_2 is required for combustion of fuel gases to occur. Nitrogen is inert. In controlled combustion systems, air and fuel must mix prior to combustion to ensure that all molecules of fuel gases are completely surrounded by oxygen. Insufficient oxygen will prevent complete combustion of the fuel.

- Mixtures with less that 4 percent natural gas in air are too lean to burn or explode.

- Mixtures of 4 - 14 percent natural gas in air can burn with a controlled flame and can explode.

- Mixtures of 15 - 100 percent natural gas in air will not explode.

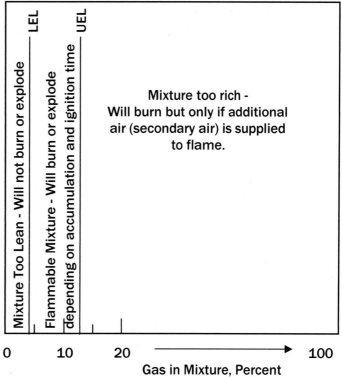

If small amounts of gas are added gradually to air, a point will be reached at which the mixture becomes explosive. The percentage of gas at this point is referenced as the Lower Explosive Limit (LEL).

If more gas is added, another point will be reached where the mixture will no longer explode. The percent gas at that point is called the Upper Explosive Limit (UEL).

Natural gas has an ignition temperature of about 1100 - 1200 degrees Fahrenheit. The ignition temperature of Propane is between 920 and 1020 degrees Fahrenheit.

Combustion

Control Fuel Gas

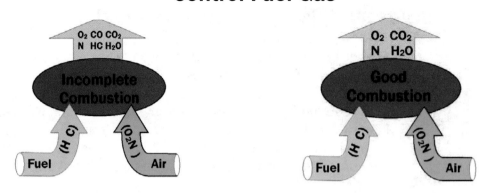

How far does fuel have to travel without leaks and without ½" WC pressure drop under full load? Have we followed Uniform Plumbing Code for sizing gas piping?

Refer to the Uniform Plumbing Code for fuel gas piping.

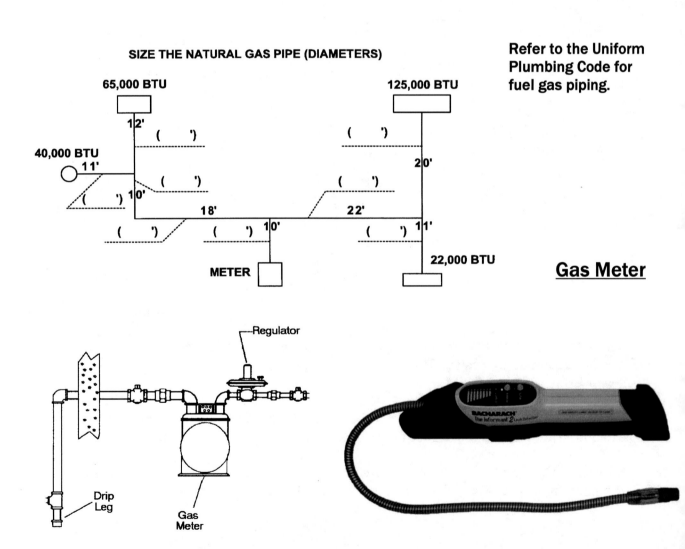

Gas Meter

For information on how to clock a gas meter, see page 67.

Combustion

Gas is delivered to appliances. We know the BTU content of our fuel. We know the input rating as listed. Fuel moves through a valve system under specific pressures. The gas moves through a gas manifold system to each individual orifice servicing each burner. Fuel pressure is measured with a U-tube manometer, the Bacharach Fyrite Pro or the Bacharach PCA (Portable Combustion analyzer) at the outlet pressure tap on the combination gas valve.

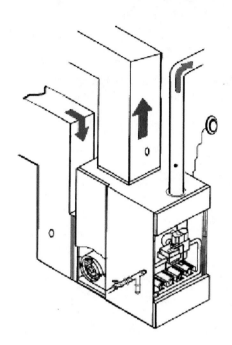
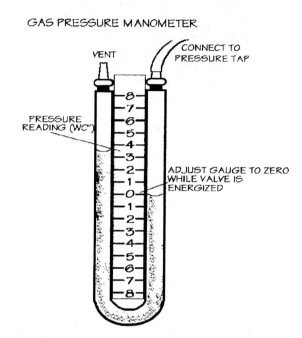

Gas leaves the manifold through an orifice, mixes with primary air and is ignited by a standing pilot or some other ignition device where a clean and properly sized burner helps maintain combustion.

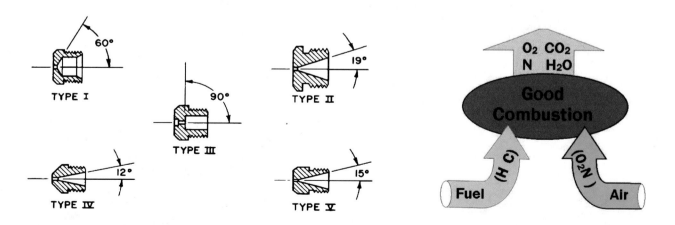

Fixed gas orifices
Reference AGA and manufacturers specifications.

Combustion

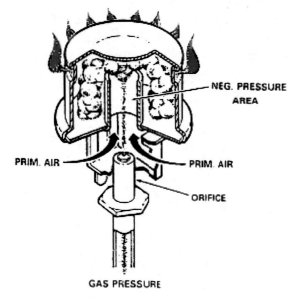

Burner with internal mixing tube (venturi)

Reference diagrams and further information
through AGA instruction material.

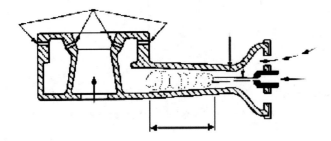

Typical Forced Air Heat Exchanger

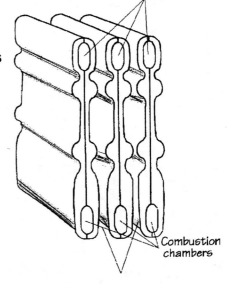

Flue gas exhaust ports

Distribution air passes through sections

Combustion

Clocking a Gas Meter

- Activate the unit being tested, making certain no other gas fired appliance is operating.

- Clock the amount of time it takes for the smallest dial to make one complete revolution.

- Using a natural gas meter timing chart (or Table XIII in the National Fuel Gas Code Book, NFPA-54, 1996), cross reference the time and appropriate dial size to determine the BTU input.

- Check and compare the calculated input with the input rating on the heating unit data plate. If the unit is under-fired or over-fired by more than 10%, check the gas pressure to the unit with a fluid filled manometer and adjust as necessary.

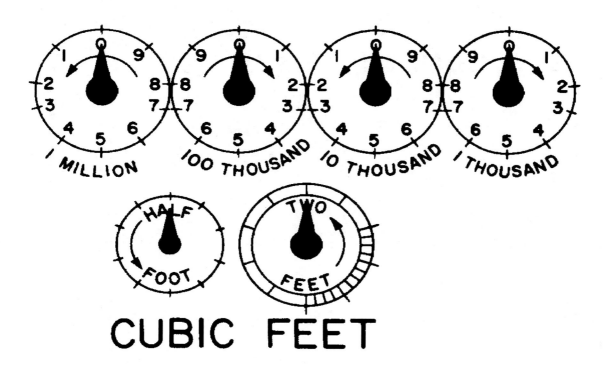

Combustion

Example:
The unit being tested takes 29 seconds for the ½ cubic foot dial to make one complete revolution. Using the chart below, this translates to 62 cubic feet per hour. Based upon the assumption that one cubic foot of natural gas has 1,000 BTU's (Check with your local utility for actual BTU content), the calculated input is 62,000 BTU's per hour.

Table XIII Gas Input to Burner in Cubic Feet per Hour

Seconds for One Revolution	Size of Test Meter Dial			Seconds for One Revolution	Size of Test Meter Dial		
	One-Half Cu Ft	One Cubic Ft per Hour	Two Cu Ft	Five Cu Ft		One-Half Cu Ft	One Cubic Ft per Hour	Two Cu Ft	Five Cu Ft

Seconds	One-Half Cu Ft	One Cu Ft	Two Cu Ft	Five Cu Ft	Seconds	One-Half Cu Ft	One Cu Ft	Two Cu Ft	Five Cu Ft
10	180	360	720	1,800	50	36	72	144	360
11	164	327	655	1,636	51	35	71	141	353
12	150	300	600	1,500	52	35	69	138	346
13	138	277	555	1,385	53	34	68	136	340
14	129	257	514	1,286	54	33	67	133	333
15	120	240	480	1,200	55	33	65	131	327
16	112	225	450	1,125	56	32	64	129	321
17	106	212	424	1,059	57	32	63	126	316
18	100	200	400	1,000	58	31	62	124	310
19	95	189	379	947	59	30	61	122	305
20	90	180	360	900	60	30	60	120	300
21	86	171	343	857	61	29	58	116	290
22	82	164	327	818	62	29	57	112	281
23	78	157	313	783	63	28	56	109	273
24	75	150	300	750	64	28	54	106	265
25	72	144	288	720	65	26	53	103	257
26	69	138	277	692	66	26	51	100	250
27	67	133	267	667	67	25	50	97	243
28	64	129	257	643	68	24	48	95	237
29	62	124	248	621	69	24	47	92	231
30	60	120	240	600	70	23	46	90	225
31	58	116	232	581	71	22	45	88	220
32	56	113	225	563	72	22	44	88	214
33	55	109	218	545	73	21	43	86	209
34	53	106	212	529	74	21	42	84	205
35	51	103	206	514	75	20	41	82	200
36	50	100	200	500	76	20	40	80	192
37	49	97	195	486	77	19	38	76	184
38	47	95	189	474	78	18	37	74	180
39	46	92	185	462	79	18	36	72	173
40	45	90	180	450	80	17	35	69	167
41	44	88	176	440	81	17	33	67	161
42	43	86	172	430	82	16	32	64	155
43	42	84	167	420	83	15	31	62	150
44	41	82	164	410	84	15	30	60	138
45	40	80	160	400	85	14	28	55	129
46	39	78	157	391	86	13	26	51	120
47	38	77	153	383	87	12	24	48	112
48	37	75	150	375	88	11	22	45	106
49	37	73	147	367	89	11	21	42	100
					180	10	20	40	100

NOTE: To convert to Btu per hour, multiply by the Btu heating value of the gas used.

Combustion

Advantages of Measuring O_2 vs CO_2

Theoretical Air Curve / Fuel Oil 2

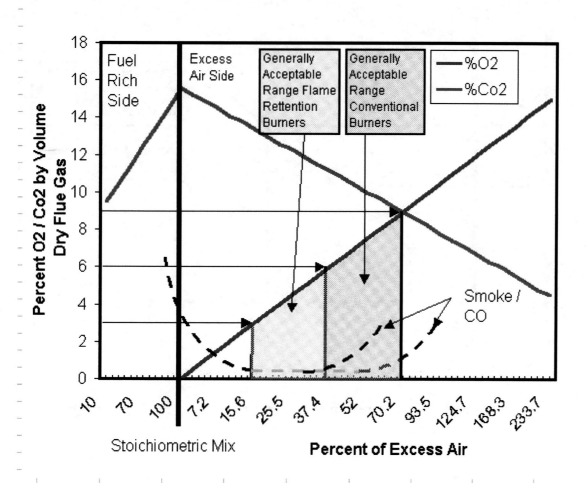

Measuring O_2 in combustion Analysis provides greater precision in establishing the relationship of O_2 / CO_2 / Excess Air, for accurate monitoring of burner performance.

Combustion

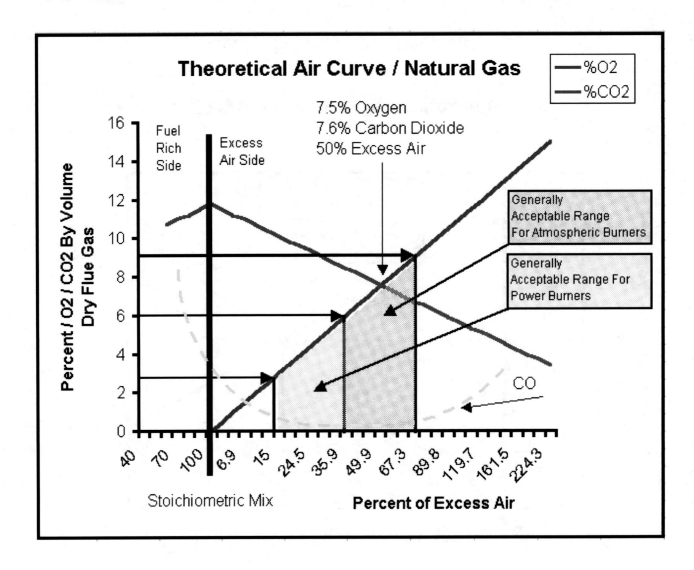

Another benefit to measuring O_2 is that CO_2 does not indicate which side of the Stoichiometric line the burner is operating on. The burner should always operate on the excess airside of the line.

Combustion

Relationship Between O_2, CO_2 and Excess Air

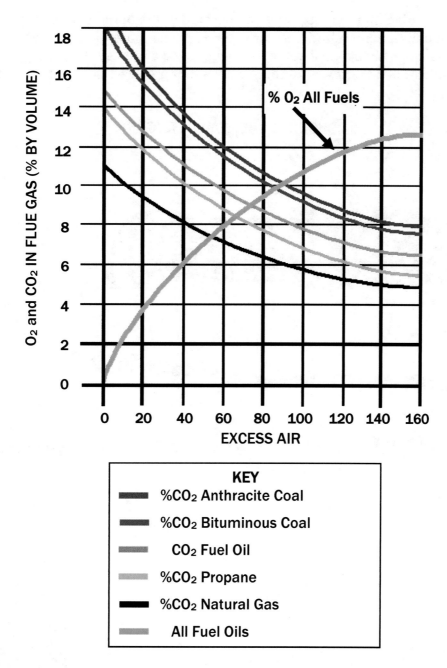

Note that CO_2 per cent is different for each fuel. O_2 per cent however, remains fairly constant with the most common fuels requiring generally a 3% to 9% O_2 content in flue gases. This is entirely dependent on the type of unit being tested. Only testing in the field under actual operating conditions can verify whether the system is installed and operating as designed.

Combustion

Leak detectors with sensors capable of detecting combustible gases like natural gas, propane and butane are used to find piping system and control failures. It is important to know the specific gravity and the BTU content of the fuel being used.

It is vital to determine that the fuel is being delivered to the burners; at proper pressures, through appropriately sized and constructed piping material, and properly sized orifices. Gas leakage should be located by an approved combustible gas detector. **Matches, candles, open flames, or other methods that provide a source of ignition should not be used.**

COMBUSTIBLE GAS LEAK DETECTORS

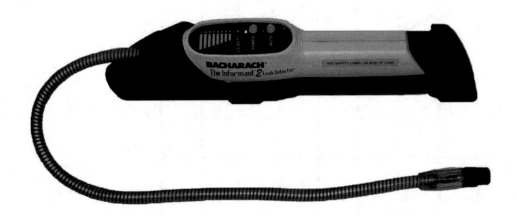

Informant 2

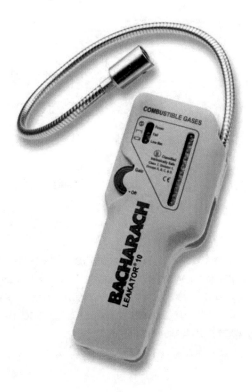

Leaktor 10

Combustion

Oil Fired Burners

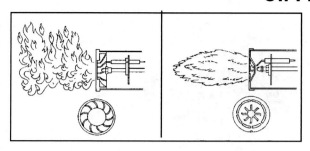

Courtesy Brookhaven National Labs

Conventional Burner   Flame Retention Burner

The amount of time oil vapor has to combust (or reside in the flame front or burning zone) has been improved dramatically with the advent of flame retention burners.

Flame retention burners violently spin the air fuel mixture. This results in better mixing and reduces the amount of excess air necessary to ensure each droplet is completely surrounded by oxygen and burns completely. As the amount of combustion air is reduced efficiency increases.

Increasing pump pressure results in finer atomization of the oil droplets. This exposes more fuel surface area to combustion air and promotes the combustion process.

To prevent over-firing the burner, the nozzle size has to be reduced using the following chart or the included formula. Also keep in mind that as pump pressure increases, so will the spray angle of the nozzle. Consequently, a nozzle with a narrower spray angle may also have to be used.

To determine fuel outputs not included on the chart, use the following formula:

New output equals (=)
Rated nozzle capacity, *times* the square root of the New pump pressure *divided* by the *rated* pump pressure (usually 100 PSI)

For example: A 1.3 GPH fuel oil nozzle with 140 PSI pump pressure.

```
New output = 1.30 X 1.40/100
           = 1.30 X 1.4
           = 1.30 X 1.18
           = 1.54 GPH
```

U.S. Gallons Per Hour - No. 2 Fuel Oil Pressure - Pounds Per Square Inch					
Rated G.P.H. @ 100 PSI	110	120	130	140
.40	.42	.44	.46	.47
.45	.47	.49	.51	.53
.50	.52	.55	.57	.59
.55	.58	.60	.62	.65
.60	.62	.65	.68	.70
.65	.68	.71	.74	.77
.70	.73	.77	.80	.83
.75	.77	.82	.86	.89
.80	.84	.88	.91	.95
.85	.89	.93	.97	1.00
.90	.94	.99	1.03	1.06
.95	1.00	1.04	1.08	1.12
1.00	1.04	1.10	1.14	1.18
1.05	1.10	1.15	1.20	1.24
1.10	1.15	1.20	1.25	1.30
1.15	1.21	1.26	1.31	1.36
1.20	1.26	1.31	1.37	1.42
1.25	1.31	1.37	1.43	1.48

Combustion

Fuel Delivery, Air, Combustion, By-product Production

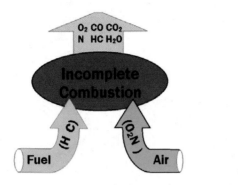
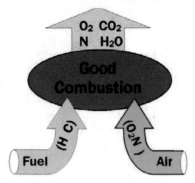

<u>Control fuel</u> - Control air fuel mixtures. Where is our air coming from? Does this <u>demand</u> for air contribute to air quality problems in the building? Control combustion. Control by-product dispersion and removal. **We have to know the fuel.**

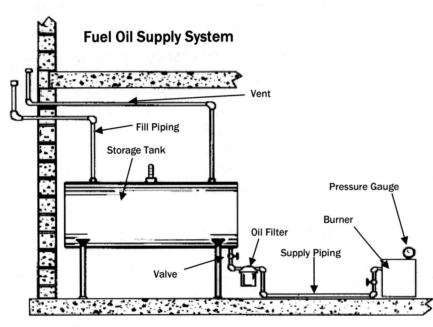

Make certain all oil line fittings are tight. Air can get sucked into leaking fittings (which may not leak oil out) with the resulting air bubbles disturbing the flame pattern as well as adding unnecessary excess air to the combustion process. <u>Always</u> use hard piped or flair connections per manufacturer instructions (PMI). **Never use compression fittings.**

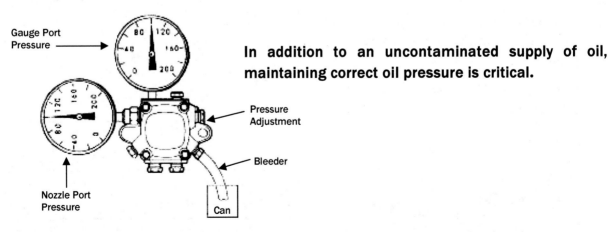

In addition to an uncontaminated supply of oil, maintaining correct oil pressure is critical.

Pressure Testing Oil Fuel Pump

Combustion

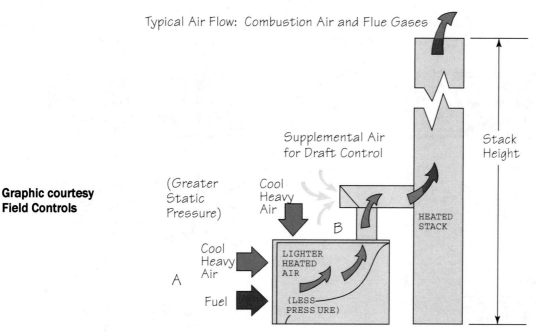

Typical Air Flow: Combustion Air and Flue Gases

Graphic courtesy Field Controls

In our attempt to keep fuel uncontaminated, under specific and steady pressures and delivered through a nozzle configuration that maximizes air fuel mixtures, we must keep in mind the effect draft has on controlling fuel.

We can only exhaust the volumes we have drawn in and mixed. Maintaining a steady draft over fire as well as in the chimney will ensure our control over our fuel. As we explore combustion further in this manual, the importance of draft measurement becomes of primary concern.

Proper sizing and adjustment of draft controls keeps our systems running cleaner and helps extend the life of combustion systems.

Draft Control; Barometric Damper

Combustion

This chart represents the products of fuel oil combustion burning with 50% excess air, which would correspond to a 7.4% O_2 in flue gases or a 10.1% CO_2 measurement.

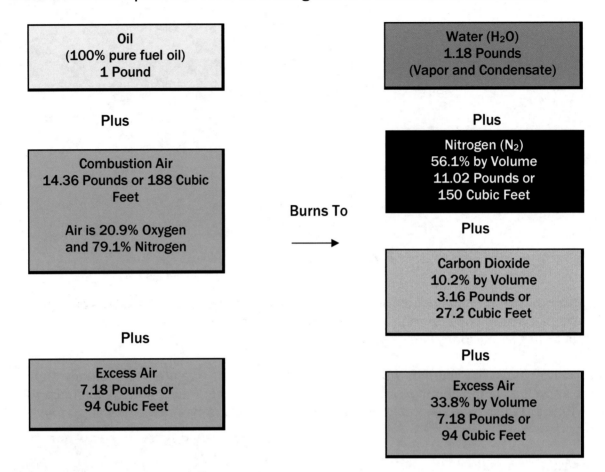

Nitrogen (N) similarly does not react during the combustion process. Nitrogen does not burn. Consequently, for every percent increase in the amount of O_2 introduced into the flame, 3.76 times as much nitrogen and volume of flue gases also enters the process. In most situations nitrogen reacts with oxygen and forms what is generally called NOx, a mixture of NO (nitric oxide) and/or NO_2 (nitrogen dioxide), both toxic emissions controlled in many areas.

Complete Combustion Needs	Complete Combustion Results
15 Cubic Feet of Air	12 Cubic Feet of Nitrogen
1 Cubic Foot of Natural Gas	2 Cubic Feet of Water vapor
	1 Cubic Foot of Carbon Dioxide
	1 Cubic Foot of Oxygen
	HEAT

While some level of excess air is necessary to ensure complete combustion, it also acts to reduce the safe efficient operation of the heating system.

Combustion

Types of Efficiencies

Combustion efficiency is a calculated measurement (in percent) of how well the heating equipment is converting a specific fuel into useable heat energy at *a specific period of time in the operation of a heating system*. Combustion test instruments evaluate these combustion gases.

Complete combustion efficiency (100%) would extract all the energy available in the fuel. However, 100% combustion efficiency is not realistically achievable due to stack loss and boiler shell losses. Various combustion processes produce efficiencies from 0% to 95+%. Combustion efficiency calculations assume complete fuel combustion and are based on three factors:

1. The chemistry of the fuel (the various proportions of hydrogen, carbon, oxygen and other compounds) and how much energy is chemically bound in the fuel.
2. The net temperature of the stack gases or how much heat is not being used.
3. The percentage of oxygen (O_2) or carbon dioxide (CO_2) by volume after the combustion process or how much O_2 did the fuel completely burn.

If your calculation shows that your equipment is losing 25% of the maximum theoretical heating value of the fuel through stack losses, your equipment is running at 75% efficiency.

Steady State Efficiency (SSE) is defined as the *point at which combustion gas content reaches equilibrium and stack temperatures stabilize*. The SSE calculated by an instrument is only reflective of fuel consumption through the heating season when O_2/CO_2, CO and stack temperatures are within the manufacturer's specifications.

Readings outside these specifications indicate that the equipment is not operating as it was engineered and designed. For example, the test results to the right show high SSE, however, the boiler is not designed to operate with a 302° stack temperature. As a result, this is a *false* SSE calculation and is actually reading Combustion Efficiency.

Thermal Efficiency reflects the rate at which heat exchange surfaces transfer heat to the transfer medium (generally water or air). Three types of heat movement impact thermal efficiency.

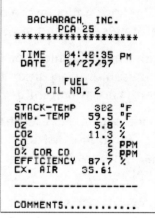
Printout of Combustion Test Results

- **Conductive/Convective heating surfaces** – also referred to as <u>secondary or indirect heating surfaces</u> including all surfaces exposed only to hot combustion gases.

- **Radiant heating surfaces** – also called direct or <u>primary heating surfaces</u>, consist of heat exchanger surfaces directly exposed to radiant heat from the flame. Radiant heat transfer is tremendously more effective than conductive/convective heat transfer and, is where most of the heat transfer occurs in a boiler or forced air system.

Seasonal Efficiency or Annual Fuel Utilization Efficiency (AFUE) is defined as actual fuel costs throughout the heating season.

Combustion

Burner Operation

All process and space heating systems are engineered and designed by the manufacturers to operate with very specific ranges of *excess combustion air*, *carbon monoxide*, *draft*, and *stack temperature* readings. Unless combustion analysis readings are within these parameters, SSE readings are 'false' and will not reflect actual consumption.

For example, an under-fired boiler with a low stack temperature may provide SSE readings that suggest efficient, economical operation. In actuality, all boilers and forced air systems are designed to operate most efficiently at their full firing rate. Under-fired burners may cause excessively low stack temperatures, which could result in condensation damage and potential flue gas spillage due to loss of stack draft.

Continual low fire (or under-fired) operation of many power burners causes the flame to burn closer to the burner head, exposing it to higher than designed temperatures and causes warping or burn off.

Looking at the flame color, shape and stability have been used as "rules of thumb" for many years but "eyeballing" will not allow you to truly optimize the safety, efficiency, full service life and environmental compliance of your equipment.

Many commercial boilers and high efficiency residential heating systems do not even have an observation port to see the flame. Even when an observation door is available, simply opening the door to view the flame changes all the actual operating conditions and characteristics of the combustion process.

Just as doctors make use of the most sophisticated instrumentation possible when diagnosing their patients, the best way to make sure that equipment you are responsible for is operating safely, and at maximum efficiency, is by using combustion instrumentation.

Traditional, chemical or Orsat type instrumentation will give you information that is comparable in accuracy to electronic instrumentation, but electronic instruments have several very important advantages.

Many electronic instruments measure on a continuous basis, like a movie or video camera. Traditional instruments are more like a still camera, which takes only one picture at time. With traditional instrumentation (the still camera) you might miss the most important picture because your camera is only capable of taking one picture at a time.

Because most electronic instruments draw flue gas samples on a continuous basis, like a video camera, you can see all of the information that will help to evaluate the operating condition of heating equipment throughout the entire cycle of operation from start up to shut down, including transient changes along the way. Electronic instruments will also do sampling and efficiency calculations rapidly and automatically.

Combustion

Some models will store and/or print out complete reports of test results or transfer the stored data to a computer while adding time and date information to the data collected.

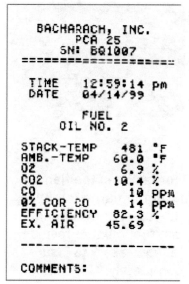

This provides hard copy documentation that the burner was operating safely and efficiently when you left the job. Undoubtedly, combustion testing will identify additional service work required. The print out can help the customer understand the nature of additional costs.

Over time, it also establishes a history of burner performance and may provide an early indication of a failing component.

A printed readout left with the customer serves as a seasonal reminder to have the burner combustion tested and also lets the customer know they have hired a company which has invested in the training and test instruments to insure safe, reliable and efficient burner operation.

To initiate the combustion process, oxygen in the combustion air and the fuel mix are ignited to produce heat. During the combustion process, carbon dioxide (CO_2) is produced in predictable quantities based on oxygen measurements and fuel types.

While the traditional, wet chemical type instrumentation determined the percentage of CO_2 in a flue gas sample, electronic instrumentation measures the amount of oxygen (O_2) remaining after the combustion process. Again, this is predictable depending on the design of the equipment. For those used to thinking in terms of CO_2, many electronic instruments provide a <u>calculated</u> CO_2 reading based upon the fuel and the O_2 percentage in the flue gases as measured by the instrument.

Also, keep in mind that O_2, CO_2 and excess air are simply different ways of conveying exactly the same information.

The air we breathe is 20.9% oxygen. As more oxygen is used to burn the fuel in the combustion process, more CO_2 is produced and diluted by excess air. Flue gas oxygen and carbon dioxide measurements are therefore inversely proportional. That is, as oxygen readings decrease, carbon dioxide readings increase.

Heating equipment is becoming more and more efficient, in part due to increased control over the amount of combustion air. In addition, residential and commercial structures are being built with much tighter envelopes in an effort to reduce fuel consumption. As a consequence, critical factors such as a sufficient combustion air supply, draft, etc. can be very easily affected by such influences as building pressure imbalances or improper fuel pressure.

Combustion

Time - Temperature - Turbulence

Combustion efficiency can be further explained in terms of the Three T's: *Time, Temperature and Turbulence.*

Time

In oil heat systems, the amount of time oil vapor has to combust (or reside in the flame front or burning zone) has been improved dramatically with the advent of flame retention burners. The primary difference between this type of burner and the older conventional style burner is that the flame retention burners violently spin the air/fuel mixture resulting in better mixing. This reduces the amount of excess combustion air necessary to insure each fuel droplet is completely surrounded by oxygen and burns completely. Efficiency increases as the amount of combustion air is reduced.

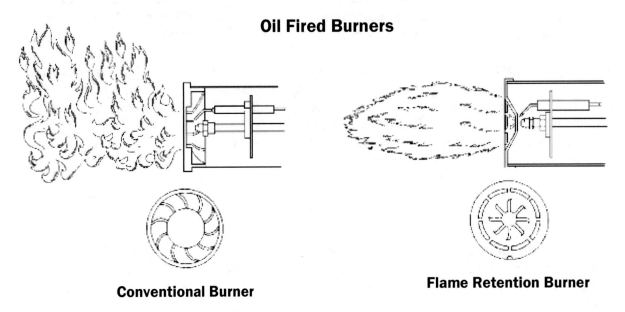

Oil Fired Burners

Conventional Burner   **Flame Retention Burner**

Courtesy Brookhaven National Labs

O_2 readings within the manufacturer's specifications document that the burner's flame retention properties are operating as it was designed and engineered.

Combustion

With gas fired systems, air and fuel mixing occurs inside the burner. Again, increasing the amount of time the fuel and air have the opportunity to thoroughly mix will help insure complete combustion.

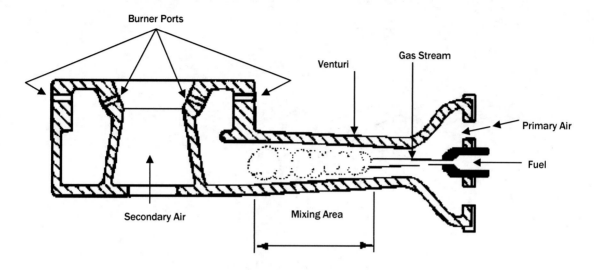

Typical Atmospheric Gas Burner

There is a common misconception that the shutter on an atmospheric burner controls combustion air intake. While some degree of combustion air does enter through the shutters, draft remains constant. As the shutters are closed, more combustion air is drawn in through the 'secondary' air intake, thus, little combustion air control is provided.

Adjustable air intake shutters on atmospheric burners primarily control the velocity at which the fuel/air mixture moves down the burner throat or 'mixing area'.

This can be demonstrated by taking a combustion test while shutters are adjusted. Unless the shutter is completely open or closed, little difference in the O_2/ CO_2 readings will be observed.

Combustion

Increasing the amount of time flue gases are in the heat exchanger also increases the amount of time for heat transfer to occur.

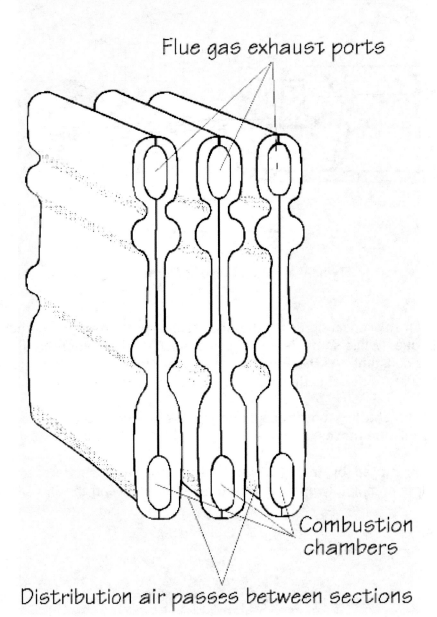

Typical Forced Air Heat Exchanger

Flue gas exhaust ports

Combustion chambers

Distribution air passes between sections

O_2 and stack temperature readings within the manufacturer's specifications document that the flue gases are moving through the heat exchanger at a velocity which allows time for maximum heat transfer while still permitting the introduction of additional combustion air.

Combustion

Temperature

As the temperature difference (DT or Delta T) between the source of heat and the material being heated increases, so does the rate of heat transfer. This heat transfer rate is measurable in forced air systems and boilers. By reducing the amount of combustion air introduced into the combustion process to the absolute minimum necessary, we increase the DT between the flame/flue gases and the distribution air or boiler water.

Radiant heat energy is much more effective in transferring heat than convective/conductive heat transfer. It is primarily put off by the flame itself and is directly related to flame temperature. As excess air is reduced, the higher flame temperature generates more radiant heat energy.

The burn rate in combustion process is very sensitive to temperature. If the flame temperature is increased by 10%, the rate of combustion more than doubles. Unfortunately, the same increase in flame temperature also increases production of NO_x gases by more than 10 times when sufficient O_2 is available.

To estimate the actual flame temperature, continue the horizontal line from the top of fuel to the point where it intersects the line corresponding to the percent of excess air. From the point of intersection, draw a vertical line up to the point where it intersects the combustion air temperature line (80°F); Then draw a horizontal line to the vertical axis – temperature rise °F.

For example:
A natural gas burner operating with 25% excess air (4.5% O_2) has an estimated flame temperature of 2930°F plus 80°F (combustion air temperature) equals 3010°F.

Note: Deduct approximately 11% for gas and 4% to 6% for oil due to water vapor in the flue gases.

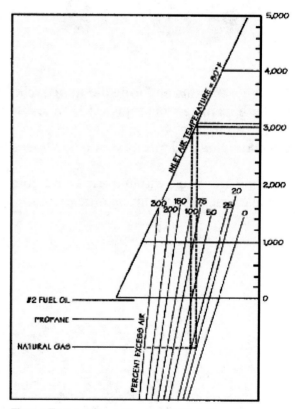

Flame Temperature/Excess Air Relationship

In the above graph, burning natural gas at 25% excess air results in a 4.5% O_2 reading and at 20% excess air results in a 4.0% reading. A ½ percent decrease in the O_2 reading represents approximately a 200°F increase in flame temperature.

O_2 readings within the manufacturer's specifications document that the flame temperature has been maximized for the most efficient radiant heat production and transfer.

Combustion

Turbulence

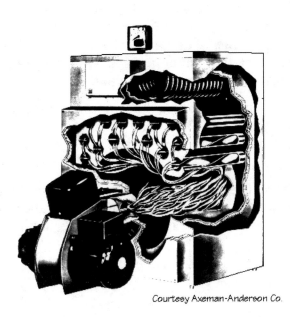

Turbulation of the fuel, air and heat source provides for more complete combustion by keeping these components in contact with each other for a longer period of time.

Agitation of flue gases in a heat exchanger provides a continual circulation of hotter flue gasses in contact with the heat exchanger surfaces. Typically, heat exchanger surfaces have a wide variety of irregular surfaces incorporating bumps, ridges etc. to provide this effect. Boilers and domestic hot water heaters commonly have turbulators that provide for this mixing process. These surfaces also produce eddy currents that recirculate flue gases and increase the amount of *time* those flue gases remain in the heat exchanger.

Courtesy Axeman-Anderson Co.

These turbulators and irregular heat exchanger surfaces are designed to be most effective at the appliance's full firing rate. Since flue gases will always take the path of least resistance through the boiler, under firing will result in a smaller volume of flue gases reducing the scrubbing and mixing of the flue gases against the fire side of the heat exchanger.

Again, O_2 readings within the manufacturer's specifications verify that the burner is firing at full capacity, maximizing efficient heat transfer.

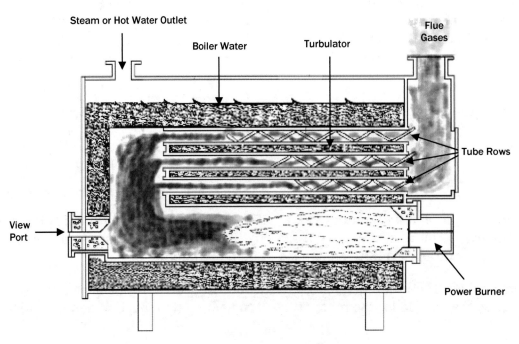

Combustion

Oxides of Nitrogen (NOx)

The primary nitrogen pollutants produced by combustion are nitric oxide (NO) and nitrogen dioxide (NO_2). These are generally referred to collectively as NO_x. Increasing evidence suggests that NOx has a direct negative effect on the human respiratory system and when exhausted into the atmosphere, reacts with moisture to produce ozone and acid rain.

For years it has been commonly accepted that NO constituted about 95% of NO_x with NO_2 making up the other 5%. More recent studies have found this may not be the case and, as such, some jurisdictions have begun to require separate measurements of NO and NO_2. Before investing in an instrument, contact your local authorities to determine which method of sampling is required. Instruments which measure both NO and NO_2 are considerably more expensive and may also require the use of a sample conditioning probe to remove moisture from the flue gas sample.

Instruments that measure NO_x generally read in Parts Per Million (PPM). Because the excess air level in the flue gases dilutes the NO_x percentage, many authorities of jurisdiction have chosen a standardized flue gas oxygen reading to which NO_x readings are corrected. For most space or process heating boilers that level is 3%.

To standardize readings from a flue gas sample use the formula:

$$\text{NOx PPM corrected to 3\% O2} = \frac{\text{Actual NO}_x \text{ PPM Reading} \times 17.9}{(20.9 - \text{Actual O}_2 \text{ Reading})}$$

Some emissions standards require levels in pounds of NO_x per million BTU's fired (Lbs. NO_x/MBtu) or other units of measurement.

POLLUTANT CONVERSIONS To convert from PPM to any of the units below: multiply PPM by the number in the correct column and row					
Fuel	Pollutant	LB/MBTU	MG/NM$_3$	MG/KG	G/GJ
Nat Gas	NOx	0.00129	2.053	20.788	0.556
Oil (#2, #6)	NOx	0.00134	2.053	24.850	0.582

Definitions (all numbers apply to values as corrected to 3% excess Oxygen)

LB/MBTU pounds of pollutants per Million BTU (British Thermal Unit)
MG/NM$_3$ milligrams of pollutant per Normal cubic meter of gas supplied
        Normal means at standard temperature and pressure
MG/KG milligrams of pollutant per Kilogram of fuel burned
G/GJ   grams of pollutant per Giga Joule (10^9 Joule)

Combustion

NOx emissions are formed in one of three ways:

- *Thermal NO_x* is produced when nitrogen and oxygen in the combustion air supply combine at high flame temperatures. Thermal NO_x is generally produced during the combustion of both gases and fuel oils.

- *Fuel NO_x* is produced when nitrogen in the fuel combines with the excess oxygen in the combustion air and is only a problem with fuel oils containing fuel bound nitrogen.

- *Prompt NO_x* is formed during the early, low temperature states of combustion and is insignificant.

NO_x control technologies vary widely depending on the required emissions standards in different jurisdictions and dictate the most cost effective strategy available for NOx reduction.

- *Reducing the amount of O_2* available to bind with nitrogen during the combustion process is probably the least expensive strategy to implement. This entails the use of a combustion analyzer to adjust the fuel/air mixture such that the amount of O_2 as measured in the flue gas sample is minimized (and still within the manufacturers' specifications). Tuning up the boiler in this manner can potentially reduce the NO_x production by as much as 10%.

- *Burning low nitrogen fuel oils* which contain approximately 18% less nitrogen can reduce NOx emissions by as much as 50%.

- *Injecting water or steam into the flame* reduces flame temperature and thus lowers overall NO_x production by as much as 80% for gas. However, this technique can result in lowering boiler efficiency by as much as 10% depending on the amount of steam or water injected. Increasing the amount of moisture in the flue gases may also lead to condensation and damage to boiler and flue passageways.

- *Flue gas re-circulation (FGR)* is one of the more commonly used methods to reduce NOx emissions and involves pulling relatively cool combustion gases from the vent system and mixing it with combustion air. Flue gases are composed of inert gases such as water vapor, carbon dioxide and nitrogen, which take heat away from the combustion process and lower flame temperatures.

Combustion

Determining the correct amount of re-circulated flue gases requires that a combustion test be performed on the boiler breech, which measures both flue gas and re-circulated flue gas O_2. Then a sample is extracted from the point at which re-circulated flue gases mix with the incoming combustion air (often referred to as the 'wind box') and an O_2 level at that point recorded. A chart available from the burner manufacturer is then used to calculate the percentage of re-circulated flue gases.

A sample of a manufacturer's chart for determining the percentage of Flue Gas Re-circulation:

O_2 Reading in Flue	6% FGR	7% FGR	8% FGR	9% FGR	10% FGR	11% FGR	12% FGR	13% FGR
				O_2 Reading in Wind box				
2.4	19.9	19.7	19.5	19.4	19.2	19.1	18.9	18.8
2.6	19.9	19.7	19.5	19.4	19.2	19.1	18.9	18.8
2.7	19.9	19.7	19.6	19.4	19.2	19.1	19.0	18.8
2.9	19.9	19.7	19.6	19.4	19.3	19.1	19.0	18.8
3.0	19.9	19.7	19.6	19.4	19.3	19.1	19.0	18.8
3.2	19.9	19.7	19.6	19.4	19.3	19.1	19.0	18.9
3.3	19.9	19.8	19.6	19.4	19.3	19.2	19.0	18.9
3.5	19.9	19.8	19.6	19.5	19.3	19.2	19.0	18.9

Flue gas re-circulation is capable of reducing NO_x emissions by as much as 75%.

Combustion

Draft

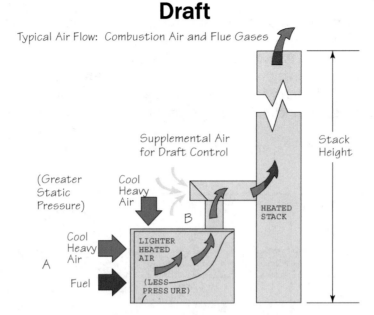

As the fuel/air mixture (A) is delivered to the boiler, it is necessary that an equal volume of flue gasses (B) enter boiler breech area. As such, the draft rate is critical in maintaining this balance. Draft diverters/hoods and barometric controls are designed to provide varying degrees of draft control and allow dilution air to mix with the flue gases to reduce the potential for condensation. A *single* acting barometric control is designed for fuel oil fired equipment while a *double* acting barometric is only approved for gas-fired systems. The double acting control door swings both directions to relieve downdrafts. Both types of barometric dampers are capable of providing a constant over fire draft, which is necessary to insure combustion air intake remains consistent under varying chimney draft conditions.

Fyrite Pro measuring draft in gas water heater

Combustion

Draft pressure is critical to the design of the particular heating system and generally falls into one of four categories:

- *Atmospheric* systems are very common and depend entirely on the slightly negative stack pressure (due to the heated flue gases being lighter than air and naturally rising) to safely exhaust flue gases to the outside, while at the same time pulling in sufficient combustion air. These systems have draft diverters or hoods located immediately downstream from the heat exchanger which allow room air to be pulled in and mixed with the products of combustion before entering the vent system.

- *Power Burners* have a mechanical blower, which delivers combustion air to the flame, but also rely on a precisely controlled overfire draft to maintain consistent combustion air intake. This generally requires the installation of a barometric control.

- *Balanced Draft* boilers, which are designed to operate under a positive pressure in the combustion chamber, generally have a breech damper (either manually or automatically controlled), which maintains the boiler combustion chamber and flue gas passageways under a positive pressure to maximize efficiency. Manufacturers' positive pressure requirements vary widely. However, *a precisely controlled negative draft in the stack is still required to remove the products of combustion at a controlled rate and to allow for the exact amount combustion air to be introduced to the flame.*

- *Forced Draft* systems also have a mechanical combustion air blower but are designed for a positive over fire pressure created, in part, by resistance to flue gas flow in the stack, which also operates under a positive pressure.

To check draft, a digital/mechanical draft gauge or inclined manometer is necessary. As with combustion testing, draft sample locations will vary depending on the type of equipment tested.

While it is of utmost importance to follow the equipment manufacturer's recommended draft readings, typical *over-fire* draft measurements are in the -.01 to -.02 Water Column Inch (WC") range on both oil and gas power burner systems.

Combustion

Typically, when a -.01 to -.02 WC" is measured over the fire, stack draft will be in the -.02 to -.04 WC" range for gas fired power burners and -.04 to -.06 for oil fired burners.

Stack draft has been commonly used to set up barometric controls and evaluate draft conditions, however, it does not necessarily guarantee correct over fire pressure which is actually the main factor influencing combustion air intake.

The combustion air intake on atmospheric equipment is so diffuse that over-fire draft readings generally cannot be obtained. Consequently, draft must be measured immediately downstream from the draft hood and should be in the -.02 to -.04 WC" range. This will insure that there is stable, continuous negative pressure in the combustion chamber to allow for the controlled introduction of air and fuel.

Draft Controls

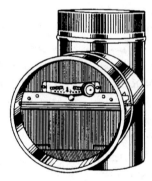

TYPE RC
Oil or Coal Residential and Commercial
Calibrated to allow easy adjustment to furnace or boiler manufacturer's specifications. Designed for settings from .02" to .08". It is so sensitive that instrumentation should be used for adjustments.

TYPE M
Oil or Coal Residential
Designed for settings from .01" to .1". Recommended for oil or coal fired residential heating applications. Features an infinitely variable screw adjustment, permitting an extremely fine instrument setting.

TYPE M + MG2
Oil, Gas, Oil/Gas, or Solid Fuels
Large Residential/Commercial
Compact, rugged, heavy duty control for any installation with 10" or larger diameter flue pipe. Adapts to any fuel. Requires only the simplest, on-the-job adjustments depending on fuel to be utilized.

Courtesy Field Controls

Combustion

Generally Acceptable Draft Measurements

Type of Heating System	Over-fire Draft	Stack Draft
Gas, Atmospheric Fan Assist (80%)	Not Applicable	-.02 to -.04 WC"
Gas, Power Burner	-.02 WC"	-.02 to -.04 WC"
Oil, Conventional	-.01 to -.02 WC"	-.02 to -.05 WC"
Oil, Flame Retention	-.005 to -.02 WC"	-.02 to -.04 WC"
Positive Over Fire Oil and Gas	+.4 to +.6 or PMI	-.02 to -.04 or PMI

Always check with the manufacturer of a particular appliance to determine the recommended over-fire/stack draft requirements.

In the past, many manufacturers recommend barometric controls be installed only when high draft conditions exist. However, field experience has shown that almost all vent systems are capable of producing excess levels of draft during certain periods of the year and that even slight variations in stack draft may influence combustion air intake.

In situations where a barometric control has been installed and subsequent testing determines high draft levels still exist, additional barometric controls can be added.

Manufacturers of barometric controls generally require that a *manual* reset spill switch be installed on the control and wired to shut the burner down in the event of an extended period of backdrafting

However, keep in mind that a spill switch will most likely only trip when a chimney is obstructed. If flue gas spillage is caused by depressurization of the mechanical room, sufficient cold air may be drawn down the stack to dilute flue gases sufficiently to not activate the spill switch.

Combustion

Best Locations

Wrong or Poor Locations

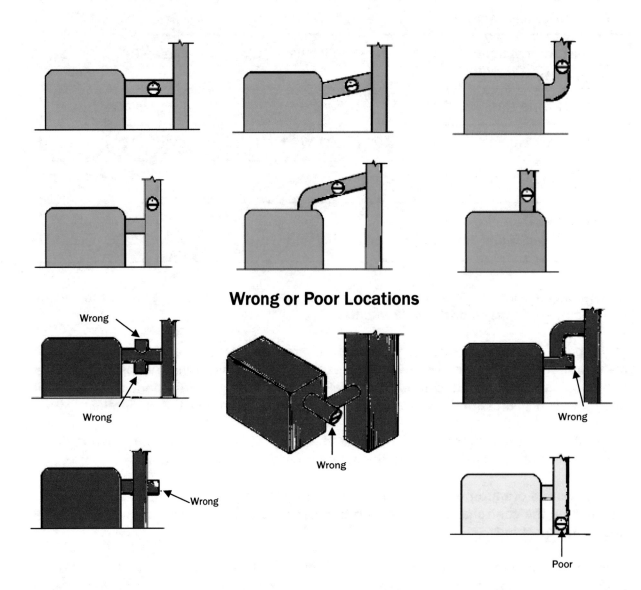

Combustion

Where multiple boilers are vented into the same chimney, attempts should be made to balance the draft of each individual boiler with separate draft controls as opposed to one draft control in the main breeching. For example, in the following diagram, draft controls should be installed in locations A or B. A draft control in location C would not provide for sufficient draft control of *each* individual boiler.

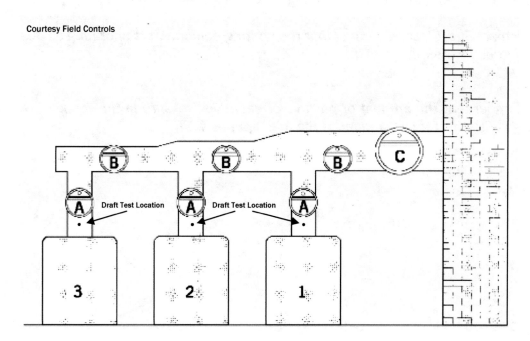

Great care must be taken to insure that the common vent functions under a wide variety of operating conditions when common venting residential fan assist (80%) furnaces or boilers with atmospheric hot water tanks.

Forced draft boilers run a positive pressure from over-fire, through to the stack termination. The entire system must be welded or otherwise sealed tightly to prevent flue gases from escaping. Also, the height of the vent termination must be limited (generally 15 feet from the breech) to prevent a negative draft. As the vent system is operating under a positive pressure, barometric controls are <u>not</u> appropriate for installation on this type system.

Taking advantage of the potential for most efficient and reliable operation requires more extensive testing be completed to properly set up this type equipment, particularly when multiple units are commonly vented. Each needs to be tested under all conceivable operating conditions.

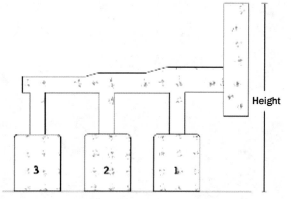

Combustion

Note: A draft reading <u>only</u> measures the difference in pressure between the inside and outside of the vent.

Combustion testing <u>verifies</u> that these gases are being drawn from the combustion chamber/heat exchanger as designed and at a sufficient rate to allow for additional introduction of the proper fuel air mixture for combustion.

Combustion and draft testing verifies that the heating equipment is operating as designed and engineered by the manufacturer.

Also, remember that the amount of positive or negative pressure in the firebox (over the fire) of a gas-fired system will influence the fuel input as well.

Combustion

Combustion Testing Procedures

To ensure safe and efficient burner operation, all residential, commercial and industrial space and process heating equipment must be properly tested for:

- Combustion Air
- Make up air
- Fuel pressure
- Carbon monoxide (CO Air Free)
- Smoke (Fuel oil only)
- Excess air (flue gas O_2)
- Stack temperature
- Draft pressure
- Temperature rise across heat exchanger
- External static pressure (on forced air systems)
- Possibly NO_x, NO, NO_2 and/or SO_2

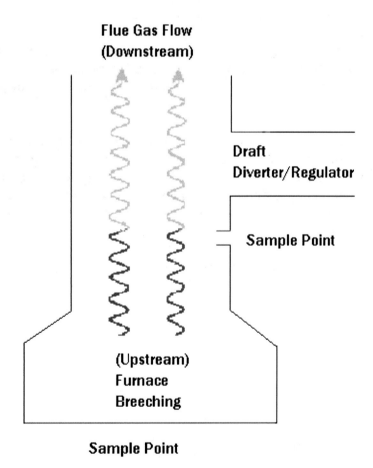

Combustion

Oxygen, Carbon Monoxide and Stack Temperature

The measurement for gases and temperature should be taken at the same point. This is normally done by selecting a sample location 'upstream' from the draft diverter/hood, barometric control or any other opening, which allows room air to enter and dilute flue gases in the stack. In larger installations it may also be necessary to extract a number of samples from inside the flue to determine the area of greatest flue gas concentration. Another common practice is to take the flue gas sample from the 'Hot Spot' or the area with the highest temperature.

Make sure that the sample point is before any draft diverter/hood or barometric damper and that the flue gasses are not diluted so the stack temperature has not been decreased by surrounding air used to balance the draft.

The sample point should also be as close to the breech area (area where the duct connects the boiler to the chimney) as possible, again, to obtain an accurate stack temperature. This may also provide a more accurate O_2 reading should air be entering the flue gas stream through joints in sheet metal vent connectors.

Oil Burners Locate the sampling hole at least six inches upstream from the breech side of the barometric control and as close to the boiler breeching as possible. In addition, the sample hole should be located twice the diameter of the pipe away from any elbows.

Gas Burners Locate the sampling hole on power burner fired boilers at least six inches upstream from the breech side of any double acting barometric control and as close to the boiler breeching as possible. Try to stay away from elbows. When testing atmospheric equipment with a draft diverter/hood, the flue gas sample should be taken inside the port(s) where flue gases exhaust the heat exchanger.

Equipment with an economizer, recouperator, or other similar device requires the sampling point be downstream from and as close as possible to the device (assuming they are installed before any draft control) to insure that the net stack temperature will provide an accurate indication of the effectiveness of the entire system.

Combustion

While combustion analysis is the emphasis here, remember that this is only one important consideration in the overall scope of HVAC system efficiency. Other conditions relating to safe, efficient and reliable heating system operation are temperature rise, duct static pressures and fuel pressures.

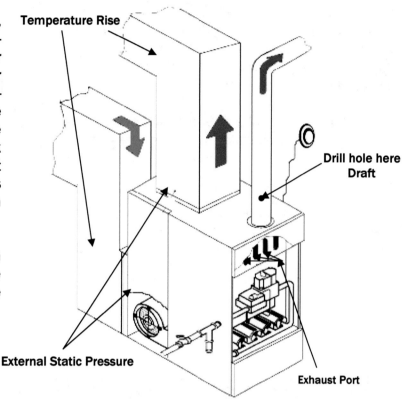

When testing atmospheric, forced air heating equipment with a clamshell or sectional heat exchanger design, test each of the exhaust ports at the top of the heat exchanger. The probe should be inserted back into each of the exhaust ports to obtain a flue gas sample before any dilution air is mixed in.

Draft tests should be taken from a hole drilled in the stack downstream from the draft hood.

Atmospheric Forced Air

Combustion and draft testing fan assist furnaces/boilers should be done through a hole drilled in the vent immediately above the inducer fan.

Combustion

Condensing furnaces/boilers can be tested through a hole drilled in the plastic vent pipe (when allowed by the manufacturer or 'local authority of jurisdiction) or taken from the exhaust termination.

In order to obtain an accurate **SSE** reading, an auxiliary thermocouple must be inserted in the combustion air intake so that a true net stack temperature is used in the calculation.

It is important to remember that the vent system on these units operates under a positive pressure. As a result, any holes in the vent need to be sealed.

If the furnace/boiler is not a 'sealed system' (which draws combustion air through a pipe from the outside) and the burners are open, CO testing should also be conducted around the burner area.

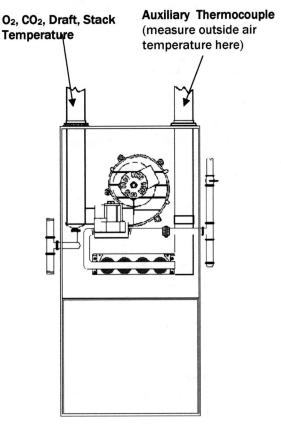

90% Condensing Furnace/Boiler

Domestic hot water heaters with the 'bell' shaped draft diverter on top can be accurately tested by attaching a section of copper tubing to the probe or using a flexible probe which is then inserted directly into the top of the fire tube below the diverter.

Another common practice is to insert the probe in the hole drilled for the draft test, direct it down and push it below the level of the draft hood.

Combustion

When testing boilers with a draft diverter mounted on the back of the equipment, flue gas samples should be taken by passing the probe from one side to the other, again upstream (toward the burner) from the opening into the draft diverter.

Draft tests should be taken from a hole drilled in the vent connector immediately above the diverter.

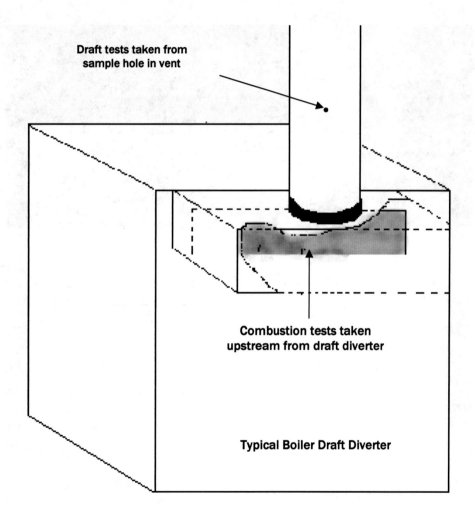

Combustion

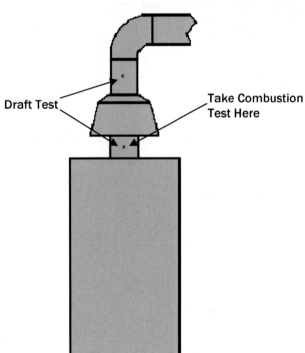

Atmospheric Boiler with Bell Shaped Draft Hood

Boilers, which have a bell shaped draft diverter directly on top, should be tested directly below the diverter through a hole drilled in the vent connector.

Should draft tests below the diverter measure insufficient draft levels, an additional test should be performed above the diverter to determine if the reason for insufficient draft is related to a chimney problem or a draft hood problem.

It is also a good idea to test any areas with openings that provide a path for combustion air to be introduced to the flame. These areas provide a path where flue gases can potentially be exhausted.

With forced air systems this area is generally located in front of the burners while many styles of boilers allow secondary combustion air to also be drawn in from all around the base of the cabinet.

Combustion

Fuel oil and gas fired power burners equipped with barometric controls require CO tests be taken at least 6" up stream from the barometric, close to the breeching.

While stack draft may be an important measurement, fuel oil and gas fired power burners require draft control over the fire to maintain a proper and controlled intake of combustion air.

Comparing stack and over-fire O_2 can verify that leakage between boiler sections, access door, etc. is minimal and the combustion test results are accurate.

Use caution when taking over fire O_2 readings. Do not expose thermocouple or sampling assembly to excess temperatures longer than necessary.

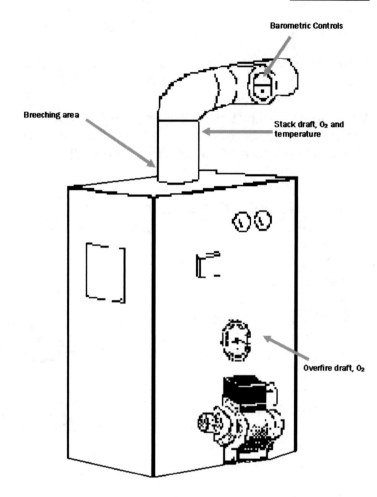

Drill hole here

When testing (primarily commercial/industrial) equipment with modulating or multiple firing rates, it is critical that tests are performed throughout the entire firing range. Typically, larger burners begin to fire at a reduced firing rate to insure a safe, reliable light off. Once ignition has been proven, air and fuel controls open to the full rated firing capacity of the boiler. Once the call for heat has been satisfied, the firing rate is slowly reduced to a minimum position before the cycle ends and the flame is extinguished.

Failing to test throughout the entire cycle of burner operation may not identify a particular point at which O_2 readings are outside the manufacturer's specifications or excess levels of CO are produced.

Combustion

Smoke Testing

Complete combustion testing of a fuel oil fired system, #1 - #6, also requires a smoke test.

When dealing with fuel oil fired heating equipment, also perform a smoke test to help identify incomplete combustion. A common misconception is that before an oil-fired appliance will produce CO, it will smoke so badly that it will be immediately evident a problem is occurring.

While it is _generally_ true that a smoky oil flame will produce CO, years of testing experience with electronic instruments has established that the reverse is not always the case.

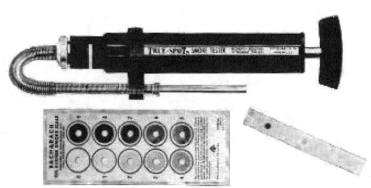

An oil-fired unit not producing a measurable amount of smoke is very capable of CO production. This is often seen when too much combustion air is introduced into the flame which results in a greater volume of flue gases being produced which acts to dilute the smoke to the point where it may not be picked up by the smoke pump filter paper.

Smoke tests are taken from the same sample location as the combustion tests. A clean piece of filter paper is inserted into the tip of the smoke tester and 10 strokes of the pump are taken.

The filter paper is removed and the dot compared to the Smoke Spot Chart.

Generally, modern flame retention burners should be set up for a zero smoke with O_2 readings within manufacturer's specifications, while an older conventional style burner may be allowed between a #1 and #2 smoke.

A "yellow" dot is an indication of unburned, raw fuel that is escaping the flame pattern and being exhausted with the flue gases.

Combustion

Acceptable Combustion Test Results

It is <u>very</u> important to consult with the manufacturer or their literature to determine acceptable ranges of O_2, CO, Stack Temperature, Steady State Efficiency, Smoke and Draft. Requirements for NOx and SO_2 emissions (if any exist) vary from local to local.

The following ranges are <u>generally</u> considered acceptable for commercial/industrial units; always check with the appliance manufacturer of specific recommendations, particularly when testing residential systems as these vary considerably from manufacturer to manufacturer.

Atmospheric Gas Fired Burners/Fan Assist

Oxygen (O_2)	7% to 9%
Stack Temperature (°F)	325° to 500°
Draft in Water Column Inches (WC")	-.02 WC" to -.04 WC" in the stack
Carbon Monoxide in Parts Per Million (PPM)	<400PPM air free (ANSI) <100PPM

Gas Fired Power Burners

Oxygen (O_2)	3% to 6%
Stack Temperature (°F)	275° to 500°
Draft in Water Column Inches (WC")	-.02 WC" Over-fire or PMI*
Carbon Monoxide in Parts Per Million (PPM)	<400ppm air free (ANSI) <100PPM

Oil Fired Power Burners

Oxygen (O_2)	4% to 7%
Stack Temperature (°F)	325° to 600°
Draft in Water Column Inches (WC")	-.02 WC" Over-fire or PMI*
Carbon Monoxide in Parts Per Million (PPM)	<100PPM air free <100PPM
Smoke	Zero or PMI

*Per Manufacturer's Instructions

Combustion

Accurate Testing

Excess air from any source (except that which is provided through the combustion air intake for complete combustion) needs to be eliminated to maximize efficiency. Just as excess combustion air reduces efficiency, air leaking in through boiler cleanout and access doors reduces the flame/flue gas temperature, and increases the volume and velocity of flue gases required in order to be vented properly.

More importantly, before a *reliable* tune up can be performed, these sources of 'unnecessary air' need to be eliminated, as combustion tests are taken downstream from the burner. This air leakage will effect the combustion test readings which are being used to determine proper fuel and air adjustments.

For example, if the burner combustion air intake is providing the proper amount of combustion air to produce clean, efficient combustion, air being drawn in through access doors, cleanout ports, etc., will increase O_2 readings on the combustion analyzer and likely result in readings that suggest the burner is operating with too much excess air. Attempts to 'fine tune' the burner by closing the combustion air intake damper or increasing fuel pressure will likely result in starving the flame for air.

Check any access doors for leakage with a smoke source to identify leakage. During service, replace any deteriorated gaskets and if necessary use high temperature silicone chalk to insure an airtight seal.

One method to determine the amount of "unnecessary" excess air from leakage in a boiler or forced air unit is to take an over-fire O_2 reading and compare with the stack O_2 reading. A higher stack O_2 reading indicates unwanted air leakage into the combustion chamber, flue passages or boiler sections. These areas must be sealed to attain accurate test results.

Eliminate "Tramp Air" leakage

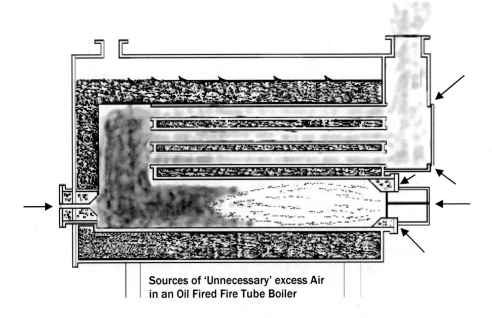

Sources of 'Unnecessary' excess Air
in an Oil Fired Fire Tube Boiler

Combustion

Proper Venting

A negative pressure switch on sidewall vented or fan assist heating equipment, only proves a certain level of negative pressure in the vent. It does not necessarily prove flue gas flow. A restriction of the combustion air supply will not necessarily cause the pressure switch to lock out. Meanwhile, inadequate combustion air may be responsible for CO production and possibly soot, both of which will increase fuel consumption and safety concerns.

Newer systems use a number of pressure switches to sense pressure drop across the heat exchanger to address this issue.

Combustion Air

While it is required to size combustion air intakes in accordance with local codes, continuous readout combustion test instruments can verify that the combustion air intakes are operating as designed. Sufficient combustion air intake can be verified by simply opening a door or window to the outside of the boiler/furnace room and noting changes in any of the readings.

Make Up Air

By the same token, observing combustion test readings while an exhaust system, air handler or clothes dryer are operated may provide information regarding the need for additional air intake to offset the indoor air removed by theses type systems.

Thermal Shock

Heat stress (thermal shock) compounds the stress of the materials. Thermal shock is one of the most common causes of boiler accidents.

As a rule of thumb, return water should never be more than 60° cooler than supply water. Generally, boiler manufacturers recommend a 20° to 40° difference, and that the burner be run at low fire or cycled for a programmed period of time before the burner brings the system to full operating capacity.

Keep this in mind when combustion testing during periods of time when the boiler hasn't been running regularly or during the summer when a chiller is in operation and there is any way for chilled water to return to the boiler.

Finally, remember other safety concerns identifiable during the course of your exposure to a particular installation are critical as well. Domestic hot water heating systems are a good example. Draft and combustion tests will help verify safe and efficient combustion. However, measuring the temperature of the water will determine the potential for scalding building occupants and is just as important.

Lowering tank temperatures have direct influences on the amount of corrosion and scale produced as well. Every 20° increase in water temperature doubles the corrosive effects of the water and increases lime scale deposits by as much as four times. This simultaneously reduces both operating efficiency and service life.

Combustion

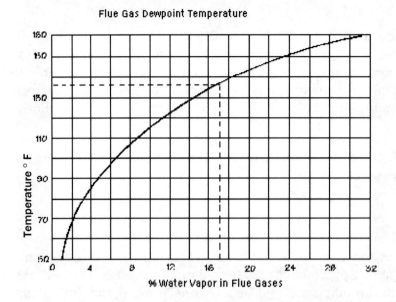

When the burner flame is adjusted to reduce the amount of excess air, the temperature at which flue gases will condense rises. To reduce the likelihood of flue gas condensation, (unless the manufacturer intentionally condenses the flue gases to derive additional heat) maintain a minimum stack temperature of 200°F above the dew point of the flue gases, and no lower than 300°F during normal operation (or per manufacturer's instructions).

To calculate the dew point temperature of flue gases, measure the O_2 or CO_2 content and determine the H_2O vapor percentage from the chart at the top.

Then, using the chart at the bottom right, follow the water vapor % value vertically to the point at which it intersects the line. Follow the line horizontally to estimate the dew point temperature.

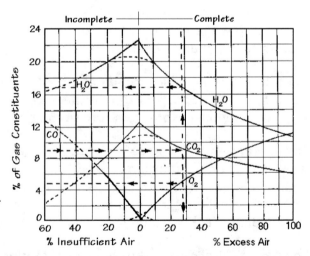

For example:

O_2 = 4.8%
CO_2 = 9%
H_2O = 16.8%

Dew Point = 137°F

As a rule of thumb, add 200° to insure the entire vent system maintains adequate temperatures. Therefore, the minimum desirable flue gas temperature would be 137°F + 200°F or 337°F.

Combustion

Boilers

A boiler is a closed vessel with a heat exchanger which uses the heat from a burner to produce hot water or steam. This heated water or steam is piped to radiators to supply space or process heating. The cooled water or steam condensate is then pumped back to the boiler where it is reheated. Generally speaking, there are two types of boiler construction.

1. Fire Tube Boilers

Fire tube boilers are generally constructed of a large steel shell, with tubes installed inside the shell so that the boiler water surrounds the tubes and the hot flue gases from the burner flow through the tubes.

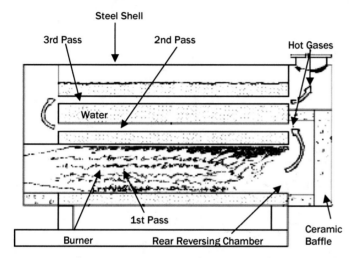

3 Pass, Dryback, Scotch Boiler

Combustion analysis can also be used to 'balance' flue gas flow through multiple pass, fire tube boilers. Flue gas samples and temperature measurements can help determine if turbulators are properly restricting the flow of flue gases and resulting in roughly an equivalent volume drawn through all rows of tubes.

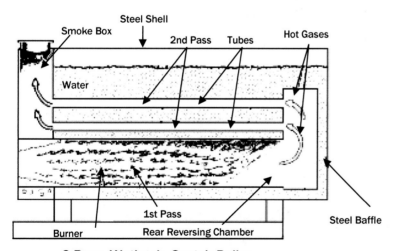

2 Pass, Wetback, Scotch Boiler

107

Combustion

2. Water Tube Boilers

A water tube boiler is constructed from one or more drums connected with tubes, which are full of water. Flue gases from the burner flow around the tubes. The pressure on the tubes is from the inside as opposed to fire tube boilers, which have the pressure on the outside. This allows the tubes to be made thinner than fire tube type boilers. It is easier to collapse a tube using external pressure then to burst a tube using internal pressure.

Some water tube boilers require the tubes be 'adjusted' as necessary. Tube spacing is designed to equally distribute the flue gas flow past all of the tubes. Should a run of tubing be slightly displaced, the larger gap between tube rows will allow a larger volume of flue gases to be pulled through, reducing the amount of total heat transfer.

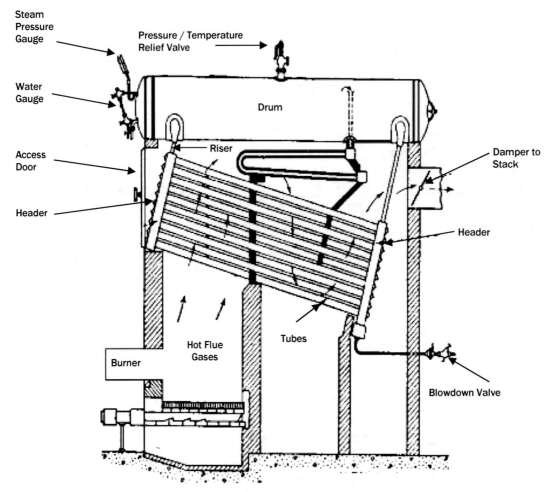

Typical Water Tube Boiler

Combustion

Stack Temperatures

Stack temperature should always be monitored and recorded on commercial power burner fired boilers. Follow "per manufacturer's instructions" which generally recommend the net stack temperature be a minimum of 100° higher than the water or steam temperature during full fire operation. Less for multiple pass boilers

This is particularly a concern with steam systems. To calculate the steam temperature, use the following chart.

Steam Pressure/Temperature Conversion Chart

PGSIG	Temperature °F
0	212°
2	218°
4	224°
6	230°
8	235°
10	240°
15	250°
20	259°
25	267°
30	274°
35	280°
40	287°
50	298°
60	307°
70	316°
80	324°
90	331°
100	338°
125	353°
150	366°

Note: Atmospheric equipment net stack temperature will generally be 50° to 100° higher than power burner applications.

Combustion

Modulating Burner Tune-up

Precise metering of the fuel and air mixtures is a basic requirement for good, clean combustion. Unfortunately, gas train valves and combustion air shutters do not allow for a 'linear' relationship between the percent of the valve opening and the volume of fuel or air passing through the valve.

As illustrated below, line 'A' represents what would be the ideal relationship between the rate of a valve opening and the amount of fuel or air flowing though the valve. Line 'B' is an example of the actual relationship. In this example, when the valve is 15% open (a typical manufacturer's recommendation for low fire), the gas flow through the valve is approaching 50%.

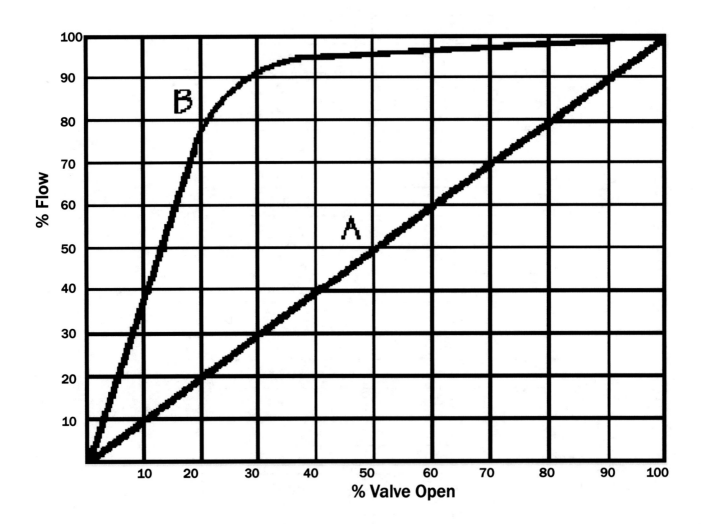

Combustion

To compensate for the lack of linear flow and provide for control over the air fuel ratio, a jackshaft assembly is used. This ties together the modulation motor, the fuel control valve and the air shutters.

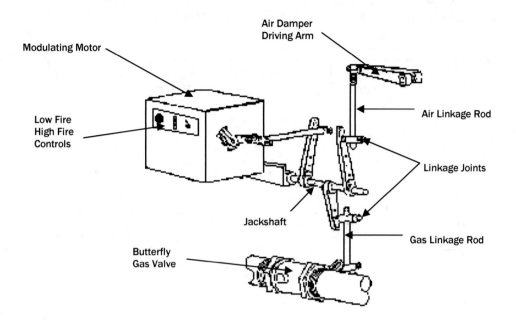

Often characterizing cams are incorporated in the control linkage to allow for even more fine-tuning of the fuel air ratio.

Given the wide variety of control arrangements available, it is critical that the manufacturer's manual is used to set up the linkage adjustments. Once linkages are set PMI, combustion testing is necessary to fine-tune the burner for the realities of the particular application.

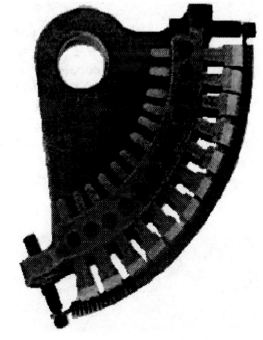

Combustion

It is also important to verify that the burner is firing to its full rated design capacity; This is sometimes performed by clocking the gas meter while the unit is firing at high fire.

While a 'fixed orifice' timing chart can be used, a correction factor has to be used to convert this reading into standard cubic feet (SCF) based upon the gas pressure at the meter.

Meter Pressure (W.C.")	Correction Factor	Meter Pressure (OSI)	Correction Factor	Meter Pressure (PSI)	Correction Factor	Meter Pressure (PSI)	Correction Factor	Meter Pressure (PSI)	Correction Factor
1	1.002	1	1.004	0.25	1.017	9	1.162	45	4.061
2	1.005	2	1.009	0.5	1.034	10	1.680	50	4.401
3	1.007	3	1.013	0.75	1.051	11	1.748	55	4.741
4	1.010	4	1.017	1	1.068	12	1.816	60	5.082
5	1.012	5	1.021	1.25	1.085	13	1.884	65	5.422
6	1.015	6	1.026	1.5	1.102	14	1.952	70	5.762
7	1.017	7	1.030	1.75	1.119	15	2.020	75	6.102
8	1.020	8	1.034	2	1.136	16	2.088	80	6.442
9	1.022	9	1.038	2.25	1.153	17	2.156	85	6.782
10	1.025	10	1.043	2.5	1.170	18	2.224	90	7.122
11	1.027	11	1.047	2.75	1.187	19	2.293	95	7.463
12	1.029	12	1.051	3	1.204	20	2.361	100	7.803

The reality of using this method to determine actual BTU input is that occasionally gas meters have been found to *not* be accurately measuring the quantity of gas passing through the meter. Additionally, often meters are not located in close proximity to the boiler room which increases the amount of time and effort required to perform this procedure.

Another method of verifying proper burner input is simply by performing a combustion test and determining that the readings are within the acceptable ranges specified by the manufacturer. If they are, the heating system is operating as designed.

Combustion

Suggested Procedure for Setting Up a Modulating Gas Fired Power Burner

1. Manually control the burner to stay at low fire, adjust barometric control and record combustion test results.
2. Verify proper gas input at low fire and adjust air linkage to comply with manufacturers' combustion specifications.
3. Manually increase the firing rate in small increments (five to ten) until full fire is attained. Record test results at each level. Also, check and record draft readings at each level.
4. Should combustion test readings drift outside the specifications, adjust fuel, combustion air linkages or barometric control to correct readings.
5. Anytime adjustments are made return to #1 and retest. Keep in mind that some compromise in ideal settings may be necessary to insure reliable performance.
6. Bring the unit up to full fire, record readings and verify acceptable ranges.
7. Perform the same procedure while ramping from high fire to low fire.
8. Once these adjustments have been made, return the control to automatic and run through (at least 10) cycles to verify proper operation.
9. Retest the boiler under varying actual load and weather conditions to monitor performance.

*Burner set up must only be performed by qualified, experienced personnel.

Combustion

Example of Manufacturers' Combustion Recommendations

Recommended Draft and Combustion Readings for Natural Gas & #2 Fuel Oil					02/19/97
Atmospheric Natural Gas Fired Boilers					
Draft (i.w.c.) Boiler Outlet	Overfire Pressure (i.w.c.)	CO_2 @ High Fire	O_2 @ High Fire	CO (ppm)	Smoke No.
-0.02 to -0.04	n/a	7.5% to 9.0%	7.5% to 4.8%	<400	0
-0.01 to -0.04	n/a	7.5% to 8.5%	7.5% to 5.8%	<400	0
-0.02 to -0.06	n/a	8.0% to 9.5%	6.7% to 4.0%	<400	0
-0.02 to -0.04	n/a	8.0% to 9.0%	6.7% to 4.8%	<400	0
-0.02 to -0.04	n/a	7.5% to 9.0%	7.5% to 4.8%	<400	0
Forced Draft Natural Gas Fired Boilers					
Draft (i.w.c.) Boiler Outlet	Overfire Pressure (i.w.c.)	CO_2 @ High Fire	O_2 @ High Fire	CO (ppm)	Smoke No.
+0.25 to -0.06	0.00 to +0.80	9.0% to 10.0%	4.8% to 3.0%	<400	0
0.00 to -0.04	-0.02 to +0.02	8.5% to 10.0%	5.8% to 3.0%	<400	0
-0.01 to -0.04	-0.02 to +0.04	7.5% to 9.5%	7.5% to 4.0%	<400	0
0.00 to -0.06	-0.02 to +0.20	8.5% to 10.0%	5.8% to 3.0%	<400	0
-0.01 to -0.04	-0.05 to +0.04	7.5% to 9.5%	7.5% to 4.8%	<400	0
0.00 to -0.06	-0.02 to +0.02	9.0% to 10.0%	4.8% to 3.0%	<400	0
0.00 to -0.06	-0.02 to +0.15	9.0% to 10.0%	4.8% to 3.0%	<400	0
+0.50 to -0.06	0.00 to +1.00	9.0% to 10.0%	4.8% to 3.0%	<400	0
+0.50 to -0.10	0.00 to +1.00	9.0% to 10.0%	4.8% to 3.0%	<400	0
+0.50 to -0.10	0.00 to +1.00	9.0% to 10.0%	4.8% to 3.0%	<400	0
Forced Draft #2 Oil Fired Boilers					
Draft (i.w.c.) Boiler Outlet	Overfire Pressure (i.w.c.)	CO_2 @ High Fire	O_2 @ High Fire	CO (ppm)	Smoke No.
+0.25 to -0.06	0.00 to +0.80	11.5% to 12.5%	5.6% to 4.2%	<400	0
0.00 to -0.04	-0.02 to +0.02	10.0% to 12.0%	7.6% to 5.0%	<400	0
-0.01 to -0.04	-0.02 to +0.04	10.0% to 12.0%	7.6% to 5.0%	<400	0
0.00 to -0.06	-0.02 to +0.02	11.5% to 12.5%	5.6% to 4.2%	<400	0
0.00 to -0.06	-0.02 to +0.15	11.5% to 12.5%	5.6% to 4.2%	<400	0
+0.50 to -0.06	0.00 to +1.00	11.5% to 12.5%	5.6% to 4.2%	<400	0
+0.50 to -0.10	0.00 to +1.00	11.5% to 12.5%	5.6% to 4.2%	<400	0
+0.50 to -0.10	0.00 to +1.00	11.5% to 12.5%	5.6% to 4.2%	<400	0

Combustion

Using the view port, located below the water connections, visually check the main burner flames at each start up after long shutdown periods or at least every six months.

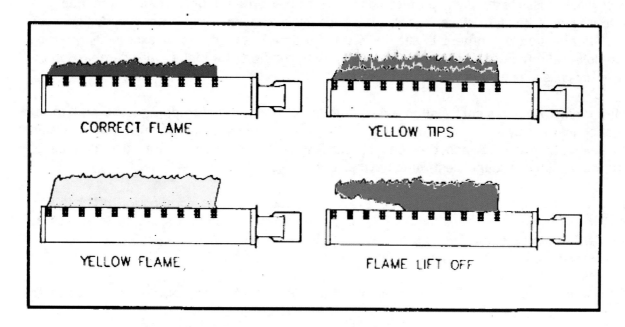

Normal Flame: A normal flame is blue, without yellow tips, with a well defined inner cone and with no flame lifting.

Yellow Tip: Yellow tip can be caused by blockage or partial obstruction of air flow to the burner(s).

Yellow Flames: Yellow flames can be caused by blockage of primary air flow to the burner(s), venturi tubes not properly in place or excessive gas input. This condition MUST be corrected immediately.

Lifting Flames: Lifting flames can be caused by over-firing the burner(s) or excessive primary air.

Flue Gas Passageways Cleaning Procedures: Any sign of soot at burners indicates a need for cleaning.

Combustion

Savings Potential

If it were possible to have perfect combustion, the amount of oxygen in the flue gas stream would be at, or close to, zero. Because perfect combustion is not practically possible, combustion equipment is set up to have a small percentage of excess O_2 present. The lower the temperature for a given O_2 (or CO_2) reading, the higher your combustion efficiency is as less heat is lost up the stack.

Fine tuning a boiler's combustion air and fuel input has a direct impact on the amount of fuel consumed by a boiler. Unfortunately, there are too many factors involved to be able to calculate exact savings that can be achieved. However, there are several rules of thumb, which can roughly estimate savings potentials.

For each 1% decrease in excess air levels introduced into the combustion process, the boiler's efficiency increases by 1/4 to 1 of a percent. While some excess air is necessary to insure complete combustion, flue gas analysis will verify that excess air is within the manufacturer's specifications and optimize efficient operation.

In a Stoichiometric Mix or "perfect" combustion, all of the fuel and oxygen introduced into the flame combine to generate <u>only</u> heat, water and carbon dioxide (CO2).

In gas-fired appliances, CO is the usual indicator of incomplete combustion. CO is well known to be a health threat, but also represents unburned fuel exhausting the appliance.

In oil-fired appliances, both CO and smoke indicate incomplete combustion. In addition to poor combustion, smoke can deposit soot on heat exchange surfaces will further reduce efficiency. Also, smoke coming out of the stack can be cause for an air quality violation and potential public relations concerns.

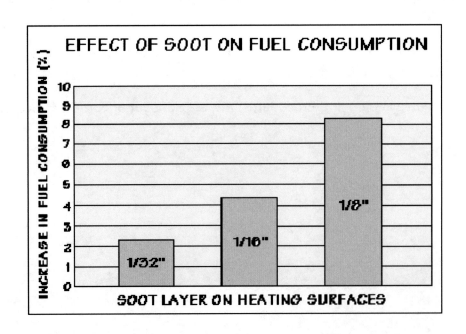

Combustion

Another method to quantify potential savings is to determine the change in SSE and use the following table to calculate fuel savings.

Savings for Every $100 Fuel Costs by Increase of Combustion Efficiency Assuming constant radiation and other unaccounted-for losses									
From an original efficiency of:	To an increased efficiency of:								
	55%	60%	65%	70%	75%	80%	85%	90%	95%
50%	$9.10	$16.70	$23.10	$28.60	$33.30	$37.50	$41.20	$44.40	$47.40
55%	---	8.30	15.15	21.50	26.70	31.20	35.30	38.90	42.10
60%	---	---	7.70	14.30	20.00	25.00	29.40	33.30	37.80
65%	---	---	---	7.10	13.30	18.80	23.50	27.80	31.60
70%	---	---	---	---	6.70	12.50	17.60	22.20	26.30
75%	---	---	---	---	---	6.30	11.80	16.70	21.10
80%	---	---	---	---	---	---	5.90	11.10	15.60
85%	---	---	---	---	---	---	---	5.60	10.50
90%	---	---	---	---	---	---	---	---	5.30

All energy conservation savings, installation costs and payback periods are estimates. Actual results may differ depending on variations in weather, usage patterns of the occupants and material and installation costs. *BACHARACH and its agents, employees, contractors and subcontractors do not guarantee the savings, contractor costs, installation costs or payback periods.*

Draft measurements verify sufficient stack draft to allow for the introduction of additional fuel air mix. Excess draft will likely pull in too much combustion air and/or excess air possibly causing the production of CO while removing hot flue gases before complete heat transfer occurs.

*A commonly accepted **rule of thumb** states that for every .01 WC" the excess draft rate can be reduced, fuel consumption will decrease 1%.*

Combustion

Combustion Troubleshooting Guide

Problem	Possible Cause	Remedy
Insufficient O_2 and/or Carbon monoxide production	Insufficient combustion air	Adjust air band setting (s)
		Check for adequate combustion air into the furnace zone
	Burner over firing	Adjust fuel input
	Low stack draft	Adjust/install barometric control
		Check for restricted heat exchanger or vent system
		Check for improperly sized or constructed chimney or vent system
High O_2 reading	Excess combustion air	Adjust air band setting (s)
	Burner under firing	Adjust fuel input
	Loose clean out ports, access doors, gasket missing in boiler sections, etc.	Repair
	Excess stack draft	Adjust/install barometric control
Fluctuation in O_2 and/or carbon monoxide readings	Changing atmospheric Conditions (i.e. Wind speed)	Evaluate for barometric control
	Cracked heat exchanger	Replace
	Loose clean out ports or gasket missing in boiler sections	Repair

(See next page for more of Combustion Troubleshooting Guide)

Combustion

Combustion Troubleshooting Guide (continued)

Problem	Possible Cause	Remedy
Excess stack temperature	Inadequate air flow across the heat exchanger	Check for dirty filter, blower and/or air conditioning coil
		Increase blower speed **Don't over amp motor!**
		Increase return or supply ducting if necessary
Insufficient stack temperature	Burner underfired	Adjust fuel input
Low temperature rise	Excess air flow past heat exchanger	Check temperature rise per manufacturer's instructions
	High fan speed	Decrease fan speed or baffle blower to reduce air flow
	Burner underfired	Adjust fuel input
Low stack draft: - less than −.02 WC" in flue gas) - less than −.04 WC" in flue (oil) - less than −.02 WC" over fire with gas or oil power burners	Improperly sized vent connector or chimney	Properly size system
	Blocked vent system	Remove blockage
	Excess elbows or long horizontal runs	Re-vent or move appliance to better location for venting
	Leakage in chimney or vent connections	Seal
	Improper vent termination	Re-vent
	Inadequate combustion air	Add combustion air
	Improperly adjusted barometric control	Adjust
High stack draft: - greater than −.04 WC" in flue (gas) - greater than −.06 WC" in flue (oil) - greater than −.02 WC" over fire with gas or oil power burners	Improper vent system sizing	Properly size system
	Absence of or improperly Adjusted barometric control	Install or adjust barometric control

Combustion

Review Questions
(answers on page 121)

1. At sea level, approximately _____ cubic feet of air is required to burn one cubic foot of natural gas at 75% efficiency.
   A. 10
   B. 15
   C. 20
   D. 25

2. **The four types of air for combustion are;**
   A. Primary, Secondary, Excess, and Draft.
   B. Primary, Secondary, Excess, and Manifold.
   C. Primary, Secondary, Excess, and Dilution.
   D. Primary, Secondary, Excess, and Ported.

3. **One British Thermal Unit (BTU) will;**
   A. raise the temperature of air 1 degree.
   B. raise the temperature of water 1 degree.
   C. raise the temperature of air 10 degrees.
   D. raise the temperature of water 10 degrees.

4. **Air is composed of approximately**
   A. 20.9% Argon and 79% Neon
   B. 20.9 % Helium and 79% Nobelium
   C. 20.9% Oxygen and 79% Nitrogen
   D. 20.9% Osmium and 79% Niobium

5. **Heat energy is expressed in BTU's. BTU is an abbreviation for:**
   A. Boron Thulium Uranium
   B. Bob's Temperature Units
   C. British Thermal Units
   D. Boron Titanium Units

6. **The ignition temperature of Natural gas is**
   A. 500 – 600 degrees Fahrenheit
   B. 1100 – 1200 degrees Fahrenheit
   C. 100 degrees Kelvin
   D. 10000 degrees Rankin

Combustion

7. An increase of pump pressure in a fuel oil system will:
   A. Decrease the life expectancy of the fuel oil nozzle
   B. Increase the volume of fuel passing through the nozzle
   C. Decrease the volume of fuel passing through the nozzle
   D. Not have an affect on the performance of fuel oil nozzles

8. One draft control is not capable of controlling draft in an installation with multiple boilers.

   True   False

9. Decreasing the O_2 reading (or increasing the CO_2 reading) by 1/2 of a percent, increases the flame temperature by approximately _____ degrees.
   A. 100
   B. 200
   C. 350
   D. 400

10. A flame with yellow tips is caused by:
    A. Overfiring
    B. Excessive gas input
    C. Airflow blockage or obstruction of burners
    D. Excessive primary air and overfiring of burners

Answers
1. B
2. C
3. B
4. B
5. C
6. C
7. B
8. T
9. B
10. C

Combustion

Pressure Measurements for Buildings; HVAC Installation, Service and Maintenance

Pressure measurements of homes and HVAC systems can help assure the following:

- All combustion appliances vent properly
- All HVAC systems are installed according to manufacturers specifications
- Buildings and their HVAC systems produce comfort
- Buildings and HVAC systems are energy efficient
- Buildings and HVAC systems are durable
- Buildings and HVAC systems do not negatively affect IAQ

Building Pressures

Reasons Why More Quality Assurance Pressure Testing is Required

More homes with more problems:

Tighter homes
- Tightness increases pressure differences within the house
- Most new homes are "tight" or under-ventilated

More exhaust fans
- Commercial quality kitchen fans
- Ventilation systems
- Higher capacity bath and laundry fans

More combustion equipment
- Fireplaces (gas and wood)
- Multiple gas water heaters and furnaces

More zones in a house
- Larger homes, more rooms
- More zones shut off by doors

More multi-story homes
- Greater stack effect
- Greater wind effect

More homes with forced air heating systems
- Increased pressure imbalances within homes

More attached garages
- Garages are tighter making them a "tight" or under-ventilated zone

Building Pressures

Societal Trends that will Increase Quality Assurance Pressure Testing

Greater awareness of carbon monoxide risks:

Transitions from prescriptive to performance based building codes
- Minnesota worst case depressurization codes
- Canadian codes (Ontario gas utilization code)

More third party efficiency interests: Commissioning
- Utilities,
- Governments
- Energy rated homes

Third party global warming interests
- Energy Star

Higher homeowner expectations of comfort
- Dual zoned cars
- "Cocooning"

Competitive pressure for service business
- Who provides the best, most thorough and measurable service to the consumer?
- More lawyers
- Quality assurance programs to lower liability rate

Building Pressures

What is Pressure?

Textbook descriptions of pressure reference it as the force pressing against a surface, stated in weight per unit area. The force acts at right angles to the surface uniformly in all directions.

Air pressure is a measurement of molecular density. The greater the molecular density, the greater the air pressure. When molecules get pushed together the pressure increases; when they get pulled apart, the density decreases. Cold air settles, pushing warmer air up.

Pressure measurements are expressed as a force exerted upon a given area and units of measurement are often referenced as Pounds per square inch (P.S.I), inches of water column (WC") and Pascals (Newtons per meter squared).

Normal air pressure at sea level is 14.7 pounds per square inch (P.S.I.). As altitude increases, the weight of the column of air decreases and a reduction of atmospheric pressure results. Atmospheric pressure at Denver, Colorado is 12.1 pounds per sq. inch (lb/sq in) and at 35,000 feet above sea level (the average flying height of a commercial let plane) is about 3.4 lb/sq in.

DRAFT

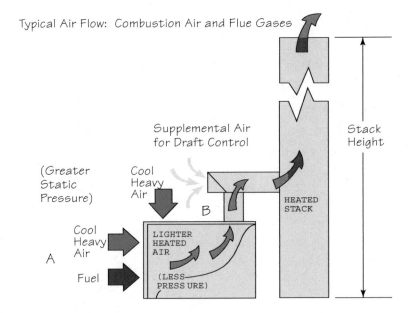

Graphic: Field Controls

When air is warmed in the combustion process, the molecules become spaced farther apart. Thus the density decreases and the colder, higher density air pushes through the venting system. This is referred to as the stack effect or as bouncy air.

Building Pressures

Warmer, Less Dense Air Rises Over Cooler, Denser Air

Which balloon has the hottest air inside?

Hot air balloons rise because of the pressure differential created by the air contained in the balloon being hotter, therefore less dense, than cooler air in the surrounding atmosphere. The balloon ascends due to its displacement of the cooler, heavier air from above. Thus, the less dense air in the balloon is forced upwards by the denser air above it just like a cork is forced upwards in water by the heavier water above it.

Air Always Moves From Areas of High Pressure to Areas of Low Pressure.

POSITIVE PRESSURE
+

NEGATIVE PRESSURE
(—)

The same is true of a fan or blower. Air on the return side of the fan is being stretched. The molecules of air are being pulled further apart, thus the air is under negative pressure and air from the rest of the house (being under higher pressure) flows towards the fan.

Building Pressures

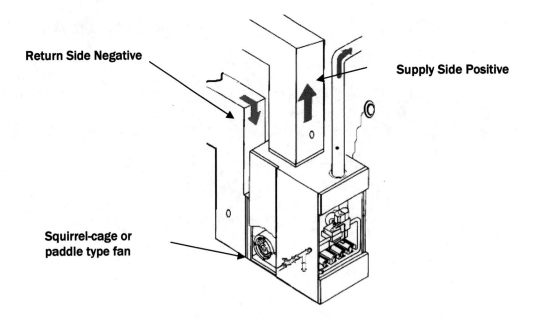

Air on the supply side of the fan is under positive pressure since the molecules of air are being pushed closer together. This air flows outward from the fan to lower pressure areas found in the house.

Fans and blowers are usually used to move air. Pressure or force is enacted with a fan, paddle wheel or squirrel cage type motion. This action causes the air to acquire a force or pressure component in a direction due to its weight and inertia.

Building Pressures

Air Flow by Building Design

Building design prescriptively or by mistake demonstrates air movements due to temperature and pressure influences.

Foul air chutes from the stall floor, up to the rooftop cupola on an old barn design demonstrates air and odor control by pressure.

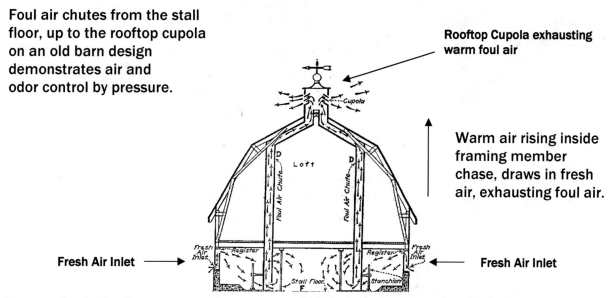

Rooftop Cupola exhausting warm foul air

Warm air rising inside framing member chase, draws in fresh air, exhausting foul air.

Fresh Air Inlet → ← **Fresh Air Inlet**

Remember! Air always moves from areas of high pressure to areas of low pressure.

In the HVAC industry, small pressures are commonly referenced in water column inches. This is done with a force pushing against a U-tube shaped container of water or same weight liquid that is open on the other end to the local atmosphere. The distance of this push is measurable in inches of water column (WC").

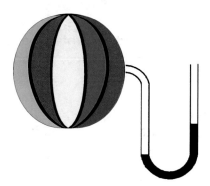

A U-tube manometer, filled with water and attached to a controlled opening on a balloon or beach ball filled with air or some other gas can measure the pressure difference between the inside and outside of the ball as the air is released, pushing against the water. If the water moves up the tube one inch in distance, the pressure is referenced as one water column inch of pressure.

Building Pressures

Types of Manometers

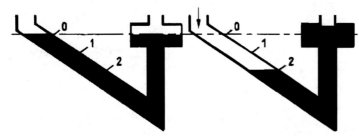

An inclined manometer is a refined version of the tube manometer. It measures smaller pressures common to air handlers and homes. This manometer has a maximum reading of one inch of water or about 249 PASCAL's.

Dial and needle manometers use a pressure-actuated diaphragm to move an indicating needle. The needle's movement is directly related to the pressure difference between the high and low pressure taps. For air flow measurement, the manometer's scale is customized to the cross sectional area of an orifice that the air is moving through (the blower door's fan housing, for example).

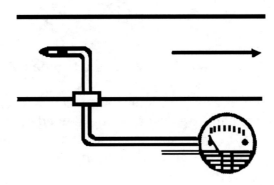

A digital Manometer is the most convenient and accurate type for diagnostic work in homes. The Bacharach Fyrite Pro 125 offers you the convenience of two point measurement on an ambient air and combustion analyzer.

This all-in-one electronic instrument has the ability to measure fuel pressure, differential pressure in WC inches or pascals, flue gas content, ambient air for CO, temperatures and it has the ability print out your results.

Building Pressures

The Manometer Law

The basis of pressure diagnostic testing is to compare the pressure in one zone to that in another zone.

Examples:
The difference in pressure of the flue compared to pressure in the mechanical room.

The difference in pressure in the mechanical room compared to outside.

The difference in pressure from one side of the air filter to the other side of the filter.

The manometer law is a simple expression that allows us to keep things straight when taking multiple pressure measurements of a building.

The law is:

_____[With reference to (WRT)]_____

For example:

The Flue (WRT) the mechanical room is (-) 5 Pascal's.

The phrase not only indicates the zone you are comparing to another zone but also the manometer configuration. The input port on the instrument would have to read the flue while the reference port would have to read the mechanical room. The first half of the expression is always connected to the input port on the instrument or manometer. The second half is always connected to the reference port.

Building Pressures

Duct Pressures

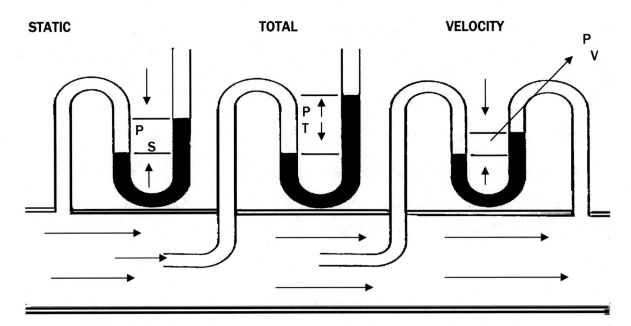

Static Pressure -
Static pressure is a bursting pressure or a pressure exerted in all directions.

Total Pressure -
Registers static and velocity pressures. Probe is positioned directly into the flow of moving air.

Velocity Pressure -
Pressure created by a moving stream of gas or air only in the direction of movement. The velocity of the air and the weight of the air create velocity pressure. The static pressure probe cancels the static pressures at the velocity pressure end and indicates true velocity pressure.

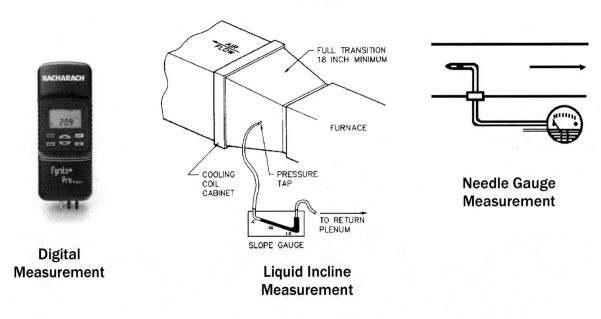

Digital Measurement   **Liquid Incline Measurement**   **Needle Gauge Measurement**

Pitot Tube

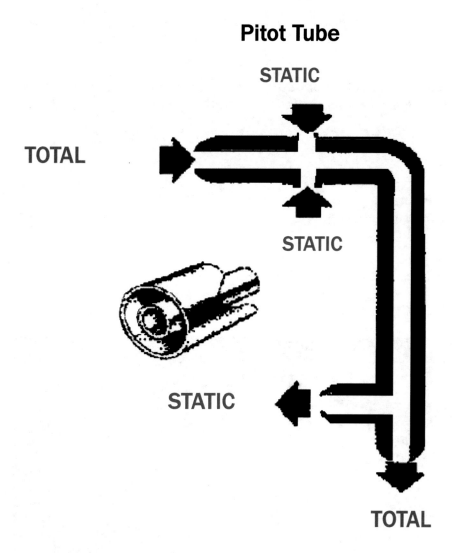

A Pitot tube consists of an impact tube (which receives total pressure input) fastened concentrically inside a second tube of slightly larger diameter which receives static pressure input from radial sensing holes around the tip. The air space between inner and outer tubes permits transfer of pressure from the sensing holes to the static pressure connection at the opposite end of the Pitot and then, through connecting tubing, to the low or negative pressure side of a manometer. When the total pressure tube is connected to the high pressure side of the manometer, velocity pressure is indicated directly.

Building Pressures

Data Plates

A gas furnace name plate, rating plate or data plate is attached to the unit and has inscribed criteria that is critical to safe installation, operation and service. This information includes:

- Manifold Pressure
- Temperature Heat Rise
- Maximum Outlet Air Temperature
- Clearances to Combustible Material
- Static Pressure
- Category Rating of the System
- (Category I, II, II or IV) and, other important information.

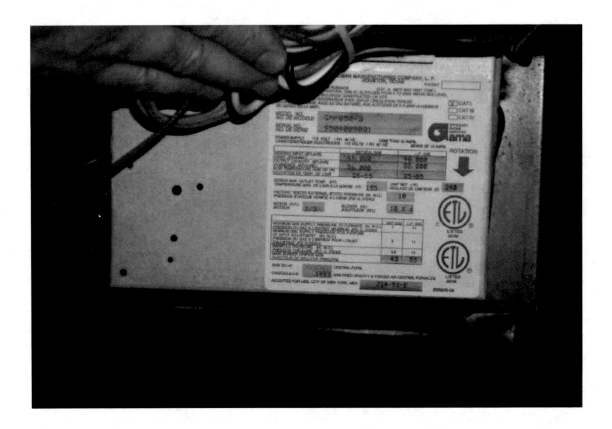

Building Pressures

Combination Gas Control

Combustion gas controls regulate gas pressure to the burner. They control the volume of gas and the amount of BTU's the burner receives. The instructions for every furnace contains information on the equipment needed to measure and adjust fuel pressure. This equipment list includes a manometer capable of reading *inches of water column*.

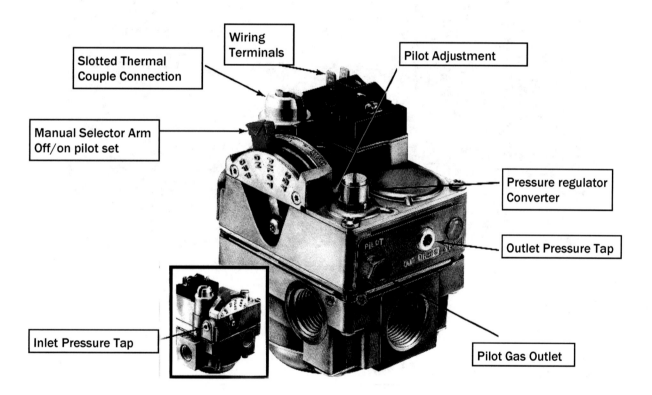

Building Pressures

External Static Pressure Testing

Use an incline manometer or Bacharach electronic instruments with differential pressure sensors to print results.

Typical Fan Curve

ESP	CFM
.1	1075
.2	1040
.3	995
.4	945
.5	895
.6	840
.7	760
.8	670

EXAMPLE:
If the external static pressure for the furnace is 150 Pascals.
150 X .004 = .6 in. WC. Thus, the fan is moving 840 CFM.

Building Pressures

Pressure Drop Across an Air Filter

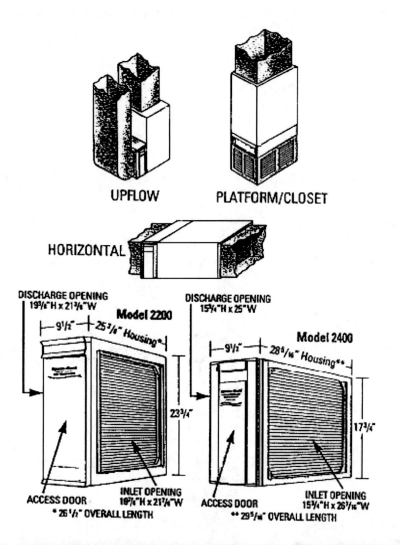

Resistance @ Airflow Volume (inches wc @ CFM)		Shipping Weight	Filtering Media Area
Model 2200	Model 2400	Model 2200	Model 2200
.05" @ 600	.06" @ 600	28 lbs.	78.6 sq. Ft.
.07" @ 800	.08" @ 800		
.09" @ 1000	.10" @ 1000	Model 2400	Model 2400
.12" @ 1200	.14" @ 1200	25 lbs.	72.3 sq. ft.
.15" @ 1400	.17" @ 1400		
.18" @ 1600	.21" @ 1600		
.21" @ 1800	.25" @ 1800		
.25" @ 2000	.29" @ 2000		

Building Pressures

Measuring Airflow by Device Static Pressure Drop

Every device produces a drop in static pressure. Knowing the drop and having the device pressure drop tables enables us to determine CFM.

Pressure Drop Method To Measure Air Flow

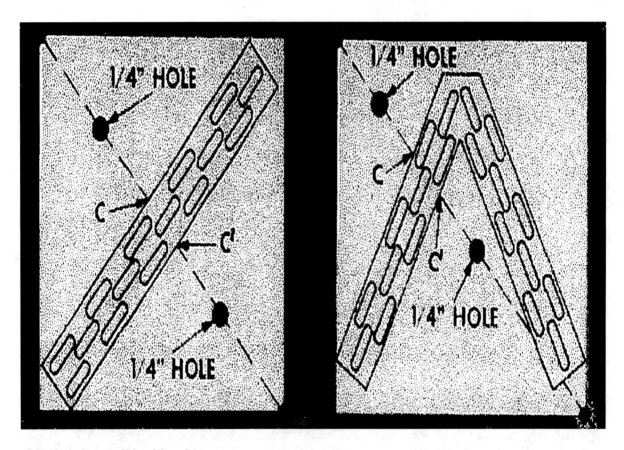

Condensing coil inside plenum

Typical Coil Pressure Drops

Building Pressures

AIRFLOW PERFORMANCE DATA (CFM vs DRY/WET STATIC @ INCHES WATER COLUMN)

Model No.	400 Dry	400 Wet	500 Dry	500 Wet	600 Dry	600 Wet	700 Dry	700 Wet	800 Dry	800 Wet	900 Dry	900 Wet	1000 Dry	1000 Wet	1100 Dry	1100 Wet	1200 Dry	1200 Wet	1300 Dry	1300 Wet	1400 Dry	1400 Wet	1500 Dry	1500 Wet	1600 Dry	1600 Wet	1700 Dry	1700 Wet	1800 Dry	1800 Wet	1900 Dry	1900 Wet	2000 Dry	2000 Wet
EPA5518SA	.06	.08	.09	.12	.12	.17	.16	.20	.20	.26																								
EPC5518BC	.06	.09	.10	.12	.13	.15	.15	.19	.19	.21																								
EX9518BC																																		
EPA5524SA					.08	.09	.11	.12	.13	.16	.17	.19	.21	.23																				
EP*5524BC			.03	.07	.06	.10	.09	.13	.12	.17	.15	.25																						
EAH5024SA																																		
EXA9524SA	.06	.09	.10	.12	.13	.15	.15	.19	.19	.21																								
EX*9524BC																																		
EPA5530BC									.13	.15	.15	.19	.18	.20	.23	.28	.27	.32																
EP*5530BC/FC									.10	.15	.13	.19	.18	.22	.21	.26	.23	.30																
EXA9530SA																																		
EX*9530FC													.20	.23	.24	.27	.27	.31	.31	.36	.36	.41												
EP*5536BC									.10	.15	.13	.19	.18	.22	.21	.26	.23	.30																
EXA9536SA											.11	.15	.14	.18	.17	.23	.20	.28	.23	.33	.27	.31												
EX*9536FC															.17	.19	.20	.23	.23	.26	.27	.30												
EP*5536FCJC																							.35	.40	.38	.44								
EAH5036SA															.16		.19		.22		.27	.30	.27	.33										
EPA5542FC																					.24	.30	.34	.38	.38	.44								
EP*5542FCJC																			.09		.30	.34	.34	.38	.13	.18	.14	.19	.17	.22				
EXA9542SA																			.11	.12	.11	.14	.11	.16	.13	.18	.16	.18	.18	.20				
EX*9542JC																									.15	.16								
EP*5548FC																					.11	.12	.13	.14										
EXA9548SA																					.21	.25	.24	.28	.26	.31								
EXA9548SB																					.16	.20												
EX*9548NC																									.23	.29	.26	.32						
EP*5548JC									.16	.19					.18	.22						.21	.26	.15	.16	.16	.18	.18	.20	.22	.24		
EP*5555NC																									.24	.25	.25	.26	.28	.30	.31	.33	.34	.36
EP*5560JC																			.12	.15	.14	.17	.15	.18	.17	.20	.19	.22	.21	.24	.22	.25	.24	.27
EP*5560NC																									.24	.25	.25	.26	.28	.30	.31	.33	.34	.36
EXA9560SA																									.12	.16	.14	.18	.16	.20	.17	.21	.20	.24
EXA9560SB																																		
EX*9560NC																																		
EAH3060SA																																		
EAH5060SA																																		

* A = Loose Coil, C = Upflow Cased Coil, H = Horizontal Cased Coil, D = Downflow Cased Coil

SPECIFICATIONS SUBJECT TO CHANGE WITHOUT NOTICE

Building Pressures

Velocity Pressure to Air Flow Calculations

For estimates of airflow, a single velocity pressure measurement of the center section of straight duct may be difficult.

First: calculate velocity V = 4,005* v pressure

Second: calculate area of duct in sq. footage
Rectangular: Area = L X W/144

Round duct: Area = $\dfrac{\pi D^2}{4 \times 144}$

Third: Calculate CFM: CFM = Area)sq. ft) X velocity

PITOT TUBE LOCATION

Duct Diameter	Area sq. ft.
4"	.0872
8"	.3488
10"	.5451
12"	.7850
14"	1.0685
16"	1.3955
18"	1.7662
20"	2.1805
22"	2.6384
24"	3.1399

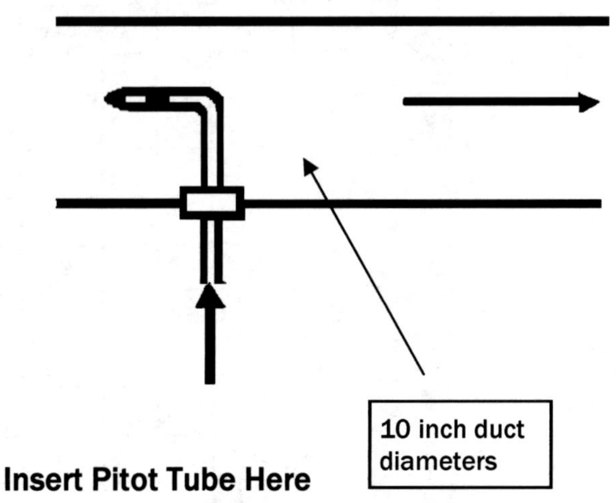

Insert Pitot Tube Here

10 inch duct diameters

Building Pressures

The Driving Forces

Driving forces are the forces that create pressure differences within homes. They are referred to as the driving forces because they drive air from one zone to another.

STACK: The house acts as a chimney

WIND: wind can play havoc with the pressures in a house. Typically, one side of the house is pressurized while the other side is depressurized.

EXHAUST FANS AND COMBUSTION VENTING: Exhaust fans and combustion appliance venting depressurizes the house as they draw air out of the house. Wood fireplaces can draw upwards of 800 CFM.

DUCT LEAKAGE: Duct leakage can pressurize and depressurize homes and zones.

SINGLE ZONE RETURN SYSTEMS AND CLOSED INTERIOR DOORS: Most new homes have only one return for the HVAC system. When interior doors are closed, rooms can become pressurized and depressurized.

Combustion air, distribution air and ventilation *air movement* are required in very specific amounts for safe, efficient operation of an HVAC system.

The forces that drive *combustion air* most often depend entirely on stack draft, a slightly negative pressure in the stack which changes in response to:

- The cycle frequency and firing rate
- Outside/inside temperature differences (DT)
- Humidity
- Winds outside the structure
- Chimney/vent diameter and construction

Distribution air is generally circulated by a motor/blower assembly, capable of forcefully moving tremendous volumes of air. The design of distribution systems is based on calculations which may not necessarily take into account such factors as:

- Unusual installation practices
- Lack of regular maintenance
- Future renovations and remodeling
- Other modifications to or changes in the structure
- Air leakage into or out of the duct system

Building Pressures

Ventilation air is required for occupant health comfort and fire safety. Ventilation air may be fan induced into the structure. More frequently, 'commonly accepted' estimates of natural air changes through air leakage in the building's thermal 'envelope' are used to base ventilation requirements.

These three systems generally operate simultaneously and each can influence operation of the other two. HVAC systems are *'listed and approved'* for particular applications, but are not always tested under actual operating conditions.

Building Pressures

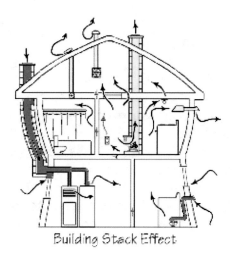
Building Stack Effect

Building Stack Effect

Buildings operate like chimneys. Warmer, less dense air rises in the structure. Air leaks out of the holes near the top of the structure (exfiltration) and leaks into the holes near the bottom of the structure (infiltration). The strength of this driving force is sufficient to back-draft combustion equipment. In most homes it is the driving force that determines the air changes per hour.

Wind

As wind blows against a side wall, higher pressures are formed. Air flows to a lower pressure (inside) through a variety of bypasses. The leeward side of the house experiences negative pressures. Combustion make-up air vents located on the leeward side of a house can act as exhaust fans depressurizing the mechanical room.

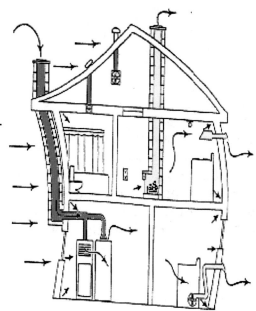

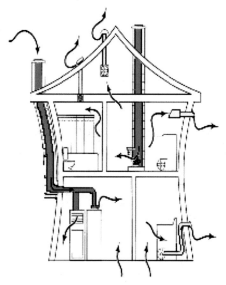

Exhaust Systems

Exhaust fans (kitchen, bath, ventilation, clothes dryers) all draw air from the house. If 100 CFM is displaced through a fan, 100 CFM must be drawn into the house. Fireplaces can act as large exhaust fans drawing up to 100 CFM out of the house.

Building Pressures

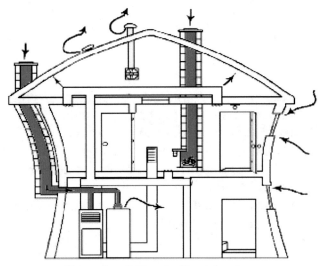

Example of Supply Duct Leakage

Supply Leakage

Supply leakage into buffer spaces such as garages, attics and crawl spaces, pressurize those zones and depressurize the entire house. The supply leaks have the same effect upon the house as exhaust fans. In both cases, air is being pushed out of the house to become depressurized.

Return Duct Leakage

Return duct leakage located in a garage, attic or crawl space draws air from these spaces, causing depressurization of these zones. This force pushes into the main body of the house causing it to become depressurized. Negative IAQ can result from drawing air from all these zones.

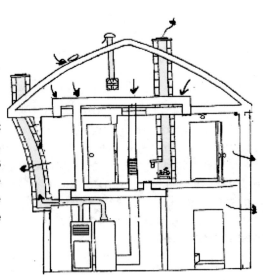

Supply Duct Leakage
(inside the same zone)

Supply duct leaks pressurize the immediate zone into which they leak. They often assist the venting system of nearby combustion appliances.

Example of Supply Duct Leakage

Building Pressures

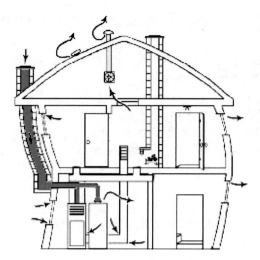

Return Duct Leakage (same zone)

Return duct leakage located in the same zone as the combustion air supply can compete with the flue draft. The return leak draws air from the combustion zone, thus depressurizing the zone and pushing it to the rest of the house, depressurizing it.

Building Pressures

Closed Door Effect in Building with Returns in One Zone

In homes without returns in every room, the potential for rooms to become pressurized and the main body of the house to become depressurized is very real.

When occupants' close interior doors with an HVAC system running, air delivered to those rooms cannot return back to the furnace. This in turn causes the rooms to be pressurized and the zone containing the return to become depressurized. Fireplaces and other vented systems are more easily reversed, and begin back drafting in these zones under this closed door effect.

Besides causing negative pressure conditions or vented system back drafting, homes utilizing forced air systems with only central returns installed do not offer airflow as prescribed by the equipment manufacturer. In the heating season this often results in some rooms not being warm enough, or possibly being too hot. In the cooling season some rooms may seem too warm when those individual rooms have their doors closed.

This problem may be causing consumers to incur additional energy costs to seek comfort in the rooms where they feel uncomfortable. Unfortunately, this issue is often overlooked by mechanical inspectors responsible for approving new installations in accordance with manufacturer specifications.

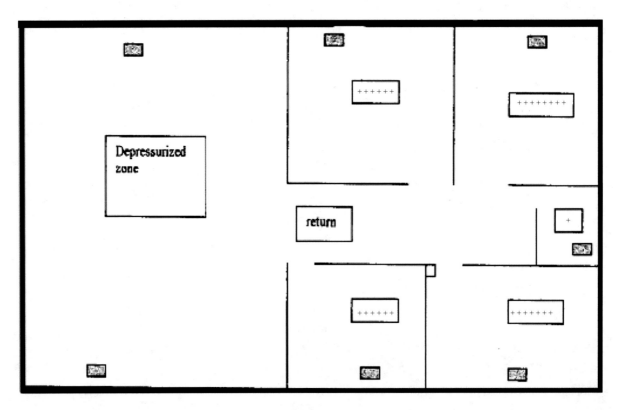

Building Pressures

Worst Case Depressurization Test

The purpose of a worst case depressurization test is to evaluate the possibility of any combustion appliance being exposed to negative pressures sufficient to cause backdrafting.

The first step is to define all the Combustion Appliance Zones, or CAZs, in the building. The CAZ is a zone where a vented combustion device is located. It may be a basement, garage, the living room or a mechanical room.

The ability of a chimney or flue to draft properly is a function of many factors. Height of the stack, outside and inside temperature, design of flue, condition of flue, other appliances sharing the same flue, wind speed and direction, tightness of the house, exhaust fans, duct leakage, location of the cold air return(s), condition of the combustion itself and homeowner interactions. In other words, there are many conditions under which a combustion appliance backdrafts or fails to vent properly. Depending on climatic conditions the worst case is likely to change on a day to day basis.

It is possible to configure most newer homes so that the negative pressure in the CAZ will exceed safe limits. Determining likely scenarios VS. extremely unlikely scenarios is the key to getting a useable test.

Worst case depressurization should not be confused with combustion by product spillage or draft testing. These are separate tests that should be conducted on all fossil fuel appliances. Because it is possible to backdraft combustion equipment during this test you must ascertain that all combustion equipment cannot come on during these tests. Backdrafting appliances can produce large amounts of CO, very quickly placing the health and safety of the technician in jeopardy. Personal CO monitors should be worn by technicians involved in furnace testing.

Building Pressures

Building Pressures

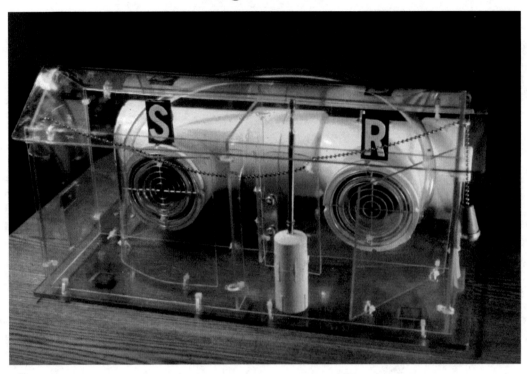

Supply Duct Leakage

Return Duct Leakage

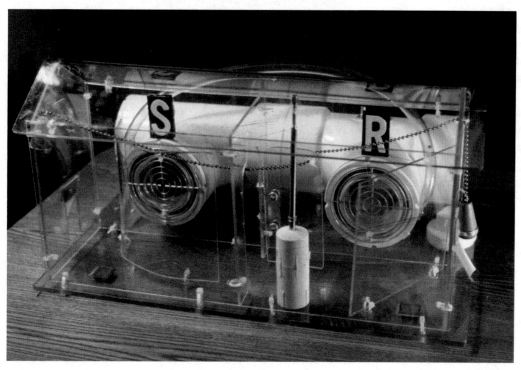

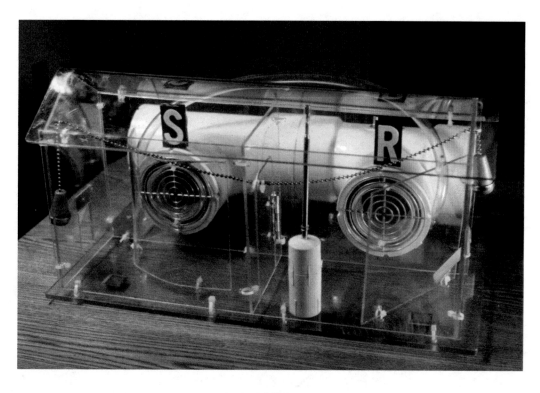

__Building Pressures__

Preparing the Building and Combustion Appliance Zone

1. Are all registers normally open or are some normally closed? Ask the occupants.

2. Are windows in rooms without returns normally open at night? If so leave them open for the test.

3. How many combustion appliances are in the same zone? Are they used at the same time? A fireplace can draw between 300 to 800 cfm. The blower door can be used to simulate a roaring fire.

4. Clean the lint trap on the clothes dryer. A dirty filter limits the cfm exhaust of the dryer.

5. Clean or replace the furnace filter. Dirty filters limit the amount of air that the HVAC fan is able to move.

6. Determine the configuration of the buffer zones. Is the garage door usually open or closed? Are any crawl space vents usually open or closed?

7. Identify the zone containing the combustion appliance. This is referred to as the CAZ. Typical appliances include gas and oil furnaces, gas and oil water heaters, gas and wood stoves, and fireplaces.

8. Identify exhaust fans (these may include but are not limited to; bath fans, kitchen fans, clothes dryers, attic fans, and exhaust fans of any kind.

Building Pressures

Doing the Test

1. Close all exterior doors and windows (with the exception of those windows normally open). If the blower door is in place, plug the fan hole (unless being used to simulate a fire in a fireplace).

2. While standing in the CAZ, place one end of an air tube outside the house, and attach the other end to the reference tap of the digital pressure gauge. An extra long hose is very handy for this test.

3. Close the interior doors to zones that do not contain return grilles.

4. Turn on the air handler (set the fan to the highest speed, in most cases this is achieved by the thermostat setting of "fan" or "air-conditioning").

5. Read pressure gauge. If gauge is reading negative the house is being depressurized. Record this number.

6. Turn on fans that are in the CAZ _____

7. Turn on fans located behind interior doors that were shut for step #3.

8. While watching the gauge, open these doors one at a time. If the CAZ goes more negative (for example minus 5 Pascals to minus 6 Pascals) keep the door open. If the CAZ becomes less negative (for example minus 5 Pascals to minus 4 Pascals) shut the door. Repeat this procedure for each room containing a fan. Record this number _____

9. Turn off the airhandler and repeat steps 8 through 10. Record this number _____

10. The highest negative number achieved is the worst case. Determine the highest negative number from line 10 or 11. Record here _____ .

Building Pressures

Setup

House Configuration (mark all appropriate conditions)

Combustion appliances in CAZ	Fireplace, woodstoves, water heater
Location of CAZ	
Interior doors closed (list rooms)	
Air handler speed	High, low, medium
Furnace filter condition	Clean / dirty
Dryer lint trap condition	Clean / dirty
Crawl space vents	Open / Closed
Garage Door	Open / Closed
Door to basement	Open / Closed
List registers closed (list rooms)	
List exhaust fans on	Kitchen, master bath, bath 1, bath 2, Jennair, Laundry, dryer, other

The backdrafting of an appliance is dependent upon many variables. Inside and outside temperatures, the type, length, and configuration of the flue, and other factors interact with the pressure inside the CAZ. Due to these interactions it is very important to determine at what pressure a given appliance will backdraft. The Canadian Mortgagee and Housing Corporation has suggested the following limits.

Appliance	Chimney Height	Unlined Chimneys on Exterior Walls	Metal Lined, Insulated or Interior Chimneys
Gas-fired furnace, boiler, DHW Heater	13 or less 14-20 21+	-5Pa -5Pa -5Pa	-5Pa -6Pa -7Pa
Oil fired furnace, boiler DHW heater	13 or less 14-20 21+	-4Pa -4Pa -4Pa	-4Pa -5Pa -6Pa
Fireplace	N/A	-3Pa	-4Pa
Induced draft appliance	N/A	-15Pa	-15Pa

Building Pressures

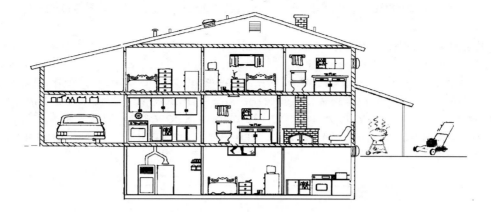

Worst Case Depressurization Testing is an important test that helps assure that backdrafting due to high negative pressures does not occur in the homes you work in. Understanding the driving forces that create negative pressures is the key to understanding backdrafting. Remember, if you don't test, you don't know.

Type of Exhaust Systems	Average Diameter of Exhaust Vent	Average Rated Flow in CFM's (cubic feet/minute)	Estimated Actual Flow in CFM's
Bathroom and kitchen range hood vents	.3"	85	53
	3 1/4 " x 10"	85	53
	4"	106	64
	7"	212	127
	8"	318	223
Clothes dryer	4"	85-127	106
Range or counter top unit with exterior vent	5"	800	300
	6"	800	500
	3 1/4 " x 10"	800	600
Wood burning fireplace			800
Standard woodstove			60
Airtight woodstove			50
Atmospheric gas or oil furnace, boiler or hot water tank	3		21
	4		38
	5		47
	6		72

Commonly Accepted Values for Exhaust Systems

Source: CMHC Chimney Safety Tests User's Manual, 2nd edition, 1998

Building Pressures

Effect of House Tightness or Zone Tightness

Tight homes (homes with relatively small areas of holes in the air barrier) are effected to a higher degree than leaky homes by driving forces. Tighter homes or zones will become more negative as a result of the same amount of air being exhausted from the zone. The chart below shows this interaction. A house with 150 square inches of holes will become negatively depressurized to minus 5 Pascals when 400 CFM fo air is being exhausted. A house with 400 square inches of holes will only be depressurized to minus 1.5 Pascals.

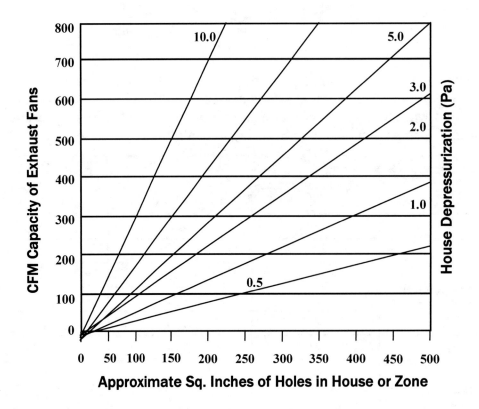

TEST, MEASURE and DOCUMENT ALL TESTS

INFORM BUILDING OCCUPANTS AND/OR AUTHORITIES HAVING JURSIDICTION OF ANY HAZARDS TO HEALTH OR LIFE.

Building Pressures

Review Questions
(answers on page 153)

1. **Air pressure is a measurement of**
   A. the temperature of the air
   B. the molecular density of the air
   C. the strength of the stack effect
   D. the concentration of combustion gases in the flue

2. **The reason heated air in a flue rises is:**
   A. heat rises
   B. the heated air contains positively charged ions which are pulled up by the negative ions in the outside air
   C. the air is less dense than the surrounding air and it displaced by the heavier surrounding air
   D. because the air is under positive pressure

3. **List four driving forces that may create pressure differences.**
   a) _____
   b) _____
   c) _____
   d) _____

4. **The small holes on the side of a Pitot tube read the:**
   A. the total pressure
   B. the static pressure
   C. the velocity pressure
   D. the head pressure

5. **Supply leakage will tend to:**
   A. Pressurize the main body of the house
   B. Depressurize the entire house
   C. Have no effect on the main body of the house since matter can neither be created or destroyed.
   D. Generate stack effect

6. **The stack effect as it applies to buildings is best defined as:**
   A. Multiple levels in homes each one being stacked on top of another
   B. The building operates like a chimney with warmer, less dense air leaving the building near the top and colder more dense air entering near the bottom of the house.
   C. What happens when the kitchen fan and bathroom fan is turned on at the same time.
   D. When more than one appliance enters the main vent or chimney on different floors of the house.

Building Pressures

7. In general a tight house will become more depressurized than a leaky house when exhaust fans are turned on.

   True   False

8. List three types of exhaust fans found in most homes:

   a) _____
   b) _____
   c) _____

9. When performing a "worst case" depressurization test, it always necessary to reference the room containing the appliance to:
   A. The master bedroom
   B. Outside
   C. The mechanical room itself
   D. The flue

10. What information is <u>not</u> found on the manufacturer rating or data plate of a furnace?
    A. Maximum outlet air temperature
    B. Vent connector diameter
    C. Category Rating of the System
    D. Temperature Heat Rise

Answers
1 2 3 4 5 6 7 8 9 10
B C B B T B B B B

Appendix A

__Appendix A__

Forms

Introduction

This Appendix is intended to provide inspectors, technicians and first responders with the forms necessary to properly assess and diagnosis situations in which CO poses a hazardous threat. These forms are designed to provide a best practices scenario for the user. Proper completion of these forms will aid in the reduction of liability and provide the information necessary for the trained professional to remediate a hazardous situation.

There are three categories of forms;

- inspector forms
- technician forms
- first responder forms.

It is the responsibility of the form user to choose the appropriate form.

Appendix A

The Bacharach Test for Carbon Monoxide

Date: _____ Test Location (ID#) _____ Phone: (___) _____

Customer Name: _____ Address _____

City _____ State _____ Zip _____ Phone: (___) _____

Service Company _____ Phone: (___) _____

Building Type _____ Reason for Inspection _____

CO Test Instrument Used _____ Last Certified Calibration Date _____

Data Logger: Time In _____ Time Out _____ Authority of Jurisdiction _____

Technician:
- Check box for each item tested or indicate: "DNT" (Did Not Test), "AT" (Additional Testing needed), or "WK" if additional worksheet used.
- Please make proper and timely referrals, note as "RF" and indicate to whom referred.
- Since it is impossible to predict every circumstance found in the field, additional documentation, testing and worksheets may be needed in some situations.

A	B	C	D	E
	Boxes checked indicate each item tested	Test & Measure	Identify & Action	Comments
	CO levels in outside air.	_____ PPM	Where?	
	CO level upon entry into building. Note rooms with increased CO levels here:	_____ PPM	Health alert? (circle) - Evacuation, 911, ventilation Safety action? (circle) - Fire Dept, Gas Co. Notes:	
	Breath analysis for CO administered for building occupants?	Occupant COHB% _____	Occupant COHB% _____	Symptoms:
	Building pressure upon entry with reference to outside.	+ or (-) _____ pascals	If yes, retest measurement: + or (-) _____ pascals	
	Open exterior windows or doors influence pressures at this entry test?	Yes ___ No ___		
	CO detector present and working?	Yes ___ No ___	Ever alarmed? Yes ___ No ___	Age & Mfg of Alarm:
	Is the installation of a CO detector/alarm recommended?	Yes ___ No ___	Brand & type recommended:	
	Working smoke alarms present?	Yes ___ No ___		

Form con't on backside →

Appendix A

Combustion System Types Present	Type of Fuel	Test hole required for CO sample?	System need cleaning or repair? Soot, scorching, corrosion? Draft measurements (WC inches) Is code inspection required?
Forced Air ____ PPM			
Oven ____ PPM			
Boiler ____ PPM			
Water Heater ____ PPM			
Fireplace ____ PPM			
Garage/Auto ____ PPM			
Other ____ PPM			
Vent connector/chimney - (Are there indications of venting problems? - ie: Insufficient/excessive draft readings, corrosion, blockage, clearances, poor installations, etc?)	Yes ____ No ____	Identify:	
Do exhaust systems (ie: kitchen/bath vents) compete with combustion air supply?	Yes ____ No ____	Identify:	
Measured building pressure imbalances? Worst case depressurization test.	Yes ____ No ____	____ pascals	Identify vented combustion systems affected by pressure. Draft pressure changes.
Do CO levels increase ambient air during testing?	Yes ____ No ____	Identify:	
Is leaky ductwork affecting vented combustion systems?	Yes ____ No ____	Note increases in CO production.	
Is there safe and sufficient combustion air supply?	Yes ____ No ____ PPM ____	Code requirement ____ Measurement ____	
Does an inspection of any system indicate any reason for replacement or immediate service?	Yes ____ No ____ N/A ____	Identify:	Left in service or shut off?
CO safety brochure given to occupants?	Yes ____ No ____	Notes:	
CO level of inside air before exiting building.	____ PPM		
CO levels of outside air after exiting building.	____ PPM		

Second opinion or further testing needed?   Yes ____   No ____   Technician's signature:
Next scheduled test or service: _____   X _____
Comments:

Appendix A

GAS FURNACE
(worksheet)

Date:_____
Phone:_____
Customer Name_____ Job #:_____
Location Address_____ Call Type:_____

Access_____Elevation_____Bldg Type_____
CO Alarms _____ Fire Alarms_____
Other Appliances_____
Furnace Mfg. _____Model #_____Serial #_____Type_____
Input Rating _____ Actual _____ Measured_____
Fuel Pressure _____Valve_____
# of Burners _____Orifice Size _____Changed To_____
Utility Co._____ Phone Number_____Fuel_____Regulator_____
Piping_____Leaks_____Dripleg_____Shutoff_____
Combustion Air Req. _____Actual _____Added_____
Venting_____
Duct System _____Filter_____
Corrections_____
Electrical Supply_____Disconnect_____Fuse_____Voltage_____
T/Stat_____Location_____Anticipator_____Setting_____Level_____
Motor _____Rating_____Amp Draw_____Speed_____
Heat Rise_____Actual_____Fan On/Off_____
Limits _____Tested_____
Corrections_____
_____

Combustion Test #_____   **Retest Results#**_____
Efficiency Rating_____% Actual_____%   Final _____% Gain _____
Carbon Monoxide (CO) _____PPM   Final CO _____PPM
Carbon Dioxide (CO2)_____%   Final CO2 _____%
Oxygen (O2)_____Blower On_____%   Final O2_____% Blower On_____
Draft Test _____Spillage_____   Draft _____Spillage_____
O.S. Temp _____I.S. Temp_____   O.S. Temp _____I.S. Temp_____
Printout_____   Printout_____
COMMENTS_____
_____
_____

Customer Education Complete_____

Customer's Signature _____Date_____
Next Scheduled Appointment _____Filter Change_____

WARNING: Your furnace has been serviced for safe and efficient operation!!
Do not attempt to repair or adjust settings!
If problems arise, contact a qualified service technician, our office, or your utility company!!!
Thank you for your help!

Appendix A

Date_____

Phone Number_____

Job Order #_____

Call Type_____

GAS WATER HEATER
(worksheet)

Customer Name_____
Location Address_____
Access_____Elevation_____Bldg Type_____

Heater Mfg._____ Model #_____ Serial #_____ Btu Rating_____
Utility Co. _____Phone Number _____Fuel _____Age of Unit_____
Gas Valve _____Setting _____Gas Leaks_____
Gas Piping _____Shutoff _____Drip Leg_____
Fuel Pressure in _____Manifold _____Adjusted To_____
Orifice Size _____Changed _____Meter Clocking_____
Water Temp _____Water Leaks _____Drain Valve_____
Pressure Relief Valve _____Piped to Drain_____
Combustion Air Req. _____Needed_____
Vent (connector) Chimney _____Changes_____

Combustion Test (initial) #_____   **Retest (final) #**_____
Carbon Monoxide (CO) _____PPM   Final CO _____PPM
Carbon Dioxide (CO2)_____%   Final CO2 _____%
Oxygen (O2)_____%   Final O2_____%
Spillage_____5 Min_____10 Min_____   Initial_____5 Min_____10 Min_____
Draft Test _____"WC   **Draft Test** _____"WC
T/Coupler Drop Out _____ Sec _____mv   _____Drop Out _____Sec _____ mv
Eco_____   _____
Burner Cleaned_____
Baffel Condition_____
WARNING LABELS_____
CORRECTIONS/COMMENTS _____
Customer Education Complete _____Technician_____
Customer's Signature _____Date_____
Next Scheduled Appointment_____

**WARNING: Your Water Heater Appliance has been serviced for safe and efficient operation!!
Do not change the temperature settings or attempt to repair !
WATER TEMPERATURES ABOVE 125° F. CAN BE DANGEROUS !**

Appendix A

Start Up Check Sheet

Dealer Information:

Name: _____

Address: _____

City _____

State _____ Zip _____

Owner Information:

Name: _____

Address: _____

City _____

State _____ Zip _____

Model# _____

Serial# _____

Type of Gas: Natural ____ LP ____

Blower Motor HP _____

Supply Voltage _____

Limits Opens at _____ (°F) or _____ (°C)

Limits Closes at _____ (°F) or _____ (°C)

Which Blower speed tap is used?

(Heating) _____

(Cooling) _____

Temperature of Supply Air

_____ (°F) or _____ (°C)

Temperature of Return Air

_____ (°F) or _____ (°C)

Rise _____ (°F) or _____ (°C)

Filter Type & Size _____

Fan "Time ON" Setting _____

Fan "Time OFF" Setting _____

Manual gas shut-off upstream of furnace/drip leg?

Yes ____ No ____

Drip-leg upstream of gas valve? Yes ____ No ____

Condensate drain connected? Yes ____ No ____

Horizontal condensate switch installed?

Yes ____ No ____

Blower speed checked? Yes ____ No ____

All electrical connections tight? Yes ____ No ____

Gas valve ok? Yes ____ No ____

Measured line pressure when firing unit: _____

Calculated firing rate: _____

Measured manifold pressure: _____

Thermostat ok? Yes ____ No ____

Sub-base ok? Yes ____ No ____

Anticipator set? Yes ____ No ____

Breaker on? Yes ____ No ____

Date of Installation: _____

Date of Start-up: _____

Technician: _____

Customer Signature: _____

Dealer Comments:

Appendix A

OIL HEATING WORKSHEET

Name _____ City _____ State _____ Zip _____ Date _____
Address _____
Telephone ( ) _____

Manufacturer _____ Model # _____
Unit Location: _____
Type: Horizontal _____ Vertical _____ System: Hydronic _____ Forced Air _____ Gravity Floor _____ Conversion _____
Up Flow _____ Down Flow _____
Rated BTU Input _____ Gun Type _____ Pot Burner _____ Rate % _____
GPM Rating: From _____ to _____

INITIAL INSPECTION & TEST

1. Oil Leaks Yes _____ No _____ Location _____
2. Line Filter Yes _____ No _____
3. Nozzle Pressure _____ Nozzle Size _____
4. Oil Pump Coupling: Good _____ Bad _____ Needs Replacing _____
5. Pump Type _____ Rotation: CW _____ CCW _____
6. Flue Condition to chimney _____
7. Amps at thermostat t _____ Anticipator Yes _____ No _____ Damper Yes _____ No _____
8. Flame characteristics at ignition: Instant _____ Delayed _____
9. Flame sensor functioning? Yes _____ No _____
10. Flue probe functioning? Yes _____ No _____
11. Transformer: Listed Voltage _____ Turn off _____
12. Delivery Fan-On Temperature _____ Fan-Off Temperature _____
13. Cycle-On limit switch Yes/No _____ Recycle Time _____
14. 5-Minute Delta T* _____ Return Air Temperature _____
15. Room Temperature _____ Heat Rise _____
16. CO Test (Undiluted flue gases) PPM measured at start-up _____ (If over 100 PPM in flue gases test room air)
17. 5-Minute Draft at Flue _____ Spillage: Yes _____ No _____
18. Steady state _____ # of pumps _____
19. Soot deposits: Heavy _____ Medium _____ Light _____
20. Gross Flue Temp _____ Combustion Air _____ Net Flue Temp _____
21. $CO_2\%$ _____ $O_2\%$ _____ Excess Air % _____ Burn Efficiency _____
22. 10-Minute Delta T _____ Return Air Temperature _____
23. Room Temperature _____ Heat Rise _____
24. 10-Min Gross Flue Temp _____ Combustion Air _____ 10-Min Burn Efficiency _____
25. $CO_2\%$ _____ $O_2\%$ _____ Setting _____ Actual Tested Temp _____
26. Limit Switch – Yes/No _____
27. Combustion Air: Interior _____ Exterior _____ Combination _____
28. Total all appliances located in common space _____ cubic feet
29. Combustion air sufficient? Yes _____ No _____
30. Return Air System: Existing Area _____ Minimum Required _____ Sq. Ft.
31. Tight _____ Leaky _____ Needs Repair _____
32. Filter Condition: Clean _____ Dirty _____ Filthy _____ Size _____
33. Delivery Air: Sufficient? Yes/No _____ Even air at registers? Yes/No _____
34. Need Repair? _____
35. Blower assembly: Okay _____ Dirty _____ Filthy _____ Vibration _____
36. Belt-driven motor _____ Belt condition _____ Belt Size _____
37. Direct Drive motor speed: Single _____ Low _____ Medium _____ High _____
38. Motor Horsepower _____ Rated Amps _____ Measured Amps _____
39. Bearing Condition _____
40. Heat Exchanger Condition: Ok _____ Dirty _____ Cracked _____
41. Can it be replaced? Yes _____ No _____

COMMENTS: _____

* Delta T is Delivery Air Temperature

RETEST & CORRECTIONS

1. Oil Leaks Repaired (how) _____
2. Line Filter: replaced _____
3. Nozzle Replaced? Yes/No _____ Type _____ Size _____ Pressure _____ Angle _____
4. Oil Pump Coupling Replaced? Yes _____ No _____
5. Pump Replaced: Yes _____ No _____ Type _____ CW _____ CCW _____
6. Flue Changes? Yes _____ No _____ Damper Installed/replaced? Yes _____ No _____
7. Thermostat: _____ adjusted _____ replaced _____
8. Flame characteristics at ignition: Instant _____ Delayed _____
9. Flame sensor replaced? Yes _____ No _____
10. Flue probe replaced? Yes _____ No _____
11. Transformer replaced? Yes/No _____ Comb Switch? Yes/No _____ Listed Voltage _____ Fan-On Temp _____ Fan-Off _____
12. New switch? Yes/No _____ Tested Limit temp _____ Limit Temp Setting _____
13. New limit switch? Yes/No _____
14. 5-Minute Delta T _____ Return Air Temperature _____
15. Ambient Air Temperature _____ 5-Minute Heat Rise _____
16. CO Test (Undiluted flue gases) PPM measured at start-up _____ Acceptable? Yes _____ No _____
17. Steady State _____ Turn off _____ # of pumps _____
18. Soot deposits: Cleaned? Yes _____ No _____
19. Gross Flue Temp _____ Combustion Air _____ Net Flue Temp _____
20. $CO_2\%$ _____ $O_2\%$ _____ Excess Air % _____ Burn Efficiency _____
21. 10-Minute Delta T _____ Return Air Temperature _____
22. Room Temperature _____ 10-Minute Heat Rise _____
23. 10-Min Gross Flue Temp _____ Combustion Air _____ 10-Min Burn Efficiency _____
24. Combustion air added? Yes/No _____ Acceptable? Yes _____ No _____
25. Limit Temperature _____ Limit Setting _____
26. If so, what size _____ Outlet size _____ Total cu. ft _____
27. Return Air Changes? Yes/No _____ What? _____
28. Sealed all leaks in Return Air system? Yes/No _____
29. Filter replaced? _____ Size _____ Modified _____
30. Delivery Air Sufficient? Yes/No _____
31. Even Delivery at registers? Yes/No _____
32. Blower cleaned? Yes/No _____ Replaced? Yes/No _____ Vibration Repaired? _____
33. Belt Replaced? Yes/No _____ Size _____ Tension _____
34. Direct Drive Motor speed _____ Motor HP _____ Rated Amps _____
35. Measured Amps _____ Replaced? Yes _____ No _____
36. Bearing replaced? _____ Bearing lubricated? _____
37. Heat Exchanger Replaced? _____ Heat Exchanger cleaned? _____

(Other changes see comments)

Inlet size _____ Outside wall _____ Horiz. _____ Pipe _____ Crawl _____ Attic _____

Appendix B

Appendix B

References

American Gas Association (AGA)
400 N Capitol Street NW
Washington DC 20001
202-824-7000
http://www.aga.org

American Society of Heating Refrigeration and Air Conditioning Engineers (ASHRAE)
1791 Tulie Circle NE
Atlanta, GA 30329

R.W. Beckett Corporation
800-645-2876

Big Sky Energy Education Services
(CO breath analysis, low level alarms)
877-213-7467 (Toll Free)
www.bigskyenergyservices.com

Boiler Efficiency Institute
PO Box 2255
Auburn, AL 36830

Brookhaven National Laboratories
PO Box 5000
Upton, NY 11973
516-344-2345

Building Performance Institute (BPI)
110 Glen St.
Glen Falls, NY 12804

Carlin Combustion Technology
800-989-2275

Comfort Company
Vic Aleshire, President
740-335-3852

Consumer Product Safety Commission
Office of Information and Public Affairs
Washington DC 20207

Jim Davis, *Carbon Monoxide Exposed*, *Combustion Fact or Fallacy*
513-226-4018
Delta T (Building Pressure Diagnostics)
Bruce Manclark
P.O. Box 11622
Harrisburg, OR 97446
541-995-6105

Mike Dina (Boiler Training)
Houston, TX
281-870-8585

Energy Conservatory
5158 Bloomington, MN 55417
612-827-1117

Environmental Protection Agency (EPA)
Washington, DC 20024

ECS Group (Energy Conservation Group)
R.W. Davis, President
740-664-5108

Field Controls
(BAROMETRIC CONTROLS, POWER VENTERS AND AGA Report # FT-C-07-93)
252-522-3031

Forbes WH, Sargent F, Foughton FJW (1945). The rate of CO uptake by normal man. American Journal of Physiology, 143:594-608

Tom Greiner (CO Case Studies)
Iowa State University Extension Service
512-294-6360

Robert Hedden
Oil Combustion
802-325-3509
Dan Holohan Associates, Inc.
800-853-8882

Inland Northwest HVAC Association
(HAZARD & WARNING TAGS)
East 811 Sprague Avenue, Suite 6
Spokane, WA 99202
800-786-3148
IAPMO (International Association of Plumbing & Mechanical Officials) Uniform Mechanical Code
1-800-854-2766

Rick Karg
Topsham, Maine
207-725-6723

Llano A Raffin T, (1990) *Management of Carbon Monoxide Poisoning*. Chest 97:165-169

National Fire Protection Association
Code references (NFPA # 54, #31, #211)
800-344-3555

Appendix B

Occupational Safety and Health Administration (OSHA)
Dept. of Labor
Washington, DC 20210

Oil Tech Talk
Alan Mercurio
570-356-2806
www.oiltechtalk.com

Positive Energy (SMOKE GENERATOR)
800-488-4340

Riello Burners
U.S. 800-992-7637
Canada, 800-387-3898

Stewart Selman
Stewart Selman & Associates
651-229-0379
www.housediagnostics.com

Southeast Technical Institute
Terry Newcomer
605-367-7628

Total Boiler Control
Ken Donithan
410-239-0308

Training and Energy Service Center
Stacy Keys
Princeton, West Virginia
304-487-6571

Underwriters Laboratories (UL)
RESIDENTIAL CARBON MONOXIDE DETECTORS
333 Pfingsten Road
Northbrook, IL 60062

University of Pennsylvania Medical Center Carbon monoxide, reference:
Thom, Ischiropoulos, Xu
215-614-0290

Appendix B

Bacharach Global Distribution

UNITED STATES

World Headquarters: Bacharach, Inc.
621 Hunt Valley Circle
New Kensington, PA 15068-7074 U.S.A.
Tel.: 724-334-5000   Fax: 724-334-5001
Toll Free in U.S.A.: 1-800-736-4666
E-mail: help@bacharach-inc.com

Bacharach Sales/Service Center - California
7281 Garden Grove Blvd., Suite H
Garden Grove, CA 92841
Tel.: 714-895-0050   Fax: 714-895-7950
E-mail: CALSERVICE@bacharach-inc.com

Bacharach Sales/Service Center - Indiana
8618 Louisiana Place
Merrillville, IN 46410
Tel.: 219-736-6178   Fax: 219-736-6269
E-mail: INDSERVICE@bacharach-inc.com

Bacharach Sales/Service Center – New Jersey
7300 Industrial Park
Rt. 130, Bldg. 22
Pennsauken, NJ 08110
Tel.: 856-665-6176   Fax: 856-665-6661
E-mail: NJSERVICE@bacharach-inc.com

Bacharach Sales/Service Center - Pennsylvania
621 Hunt Valley Circle
New Kensington, PA 15068-7074
Tel.: 724-334-5051   Fax: 724-334-5723
E-mail: help@bacharach-inc.com

Bacharach Sales/Service Center - Texas
5151 Mitchelldale, B-4
Houston, TX 77092
Tel.: 713-683-8141   Fax: 713-683-9437
E-mail: TXSERVICE@bacharach-inc.com

CANADA
Bacharach of Canada Sales/Service Center - Ontario
250 Shields Court Unit #3
Markham, Ontario L3R 9W7, Canada
Tel.: 905-470-8985   Fax: 905-470-8963
Toll Free in Canada: 800-328-5217
E-mail: bachcan@idirect.com

MEXICO

Bacharach de México S.de R. L. de C.V. Sales/ Service Center
Playa Regatas No. 473 Tercer Piso
Col. Militar Marte
Delegación Iztacalco, 08830
México D.F., México
Tels.:   +52-55-56-34-77-40 / 41
Fax: +52-55-56-34-77-38
E-mail: bacharach@prodigy.net.mx

BRAZIL

Bacharach International
Rua das las Laranjeiras, 498/602
Rio de Janeiro, RJ 22.240-002, Brazil
Phone : +55-21-2556-1156
Fax   :   +55-21-2557-9937
Cell. :   +55-21-9976-5088
E-mail : bacharach@samerica.com

EUROPE

European Headquarters
Bacharach Instruments
Sovereign House
Queensway, Lemington Spa
Warwickshire, England CV31 3JR
Tel.: +44-1926-338111
Fax: +44-1926-338110
E-mail: sales@bacharach-europe.com
Website: www.bacharach-europe.com

Sales / Service Center - Denmark
Bacharach Instruments Int'l
P.O. Box 44
39 Lindegade
DK 6070 Christiansfeld, Denmark
Tel.: +45-74-563171
Fax: +45-74-563178
E-mail: mail@bacharach.dk